Peter Scholz

Inductively Coupled Antenna Systems

Peter Scholz

Inductively Coupled Antenna Systems

Analysis and Numerical Modeling

Südwestdeutscher Verlag für Hochschulschriften

Imprint
Any brand names and product names mentioned in this book are subject to trademark, brand or patent protection and are trademarks or registered trademarks of their respective holders. The use of brand names, product names, common names, trade names, product descriptions etc. even without a particular marking in this work is in no way to be construed to mean that such names may be regarded as unrestricted in respect of trademark and brand protection legislation and could thus be used by anyone.

Publisher:
Südwestdeutscher Verlag für Hochschulschriften
is a trademark of
Dodo Books Indian Ocean Ltd., member of the OmniScriptum S.R.L Publishing group
str. A.Russo 15, of. 61, Chisinau-2068, Republic of Moldova Europe
Printed at: see last page
ISBN: 978-3-8381-2325-7

Zugl. / Approved by: Darmstadt, TU, Diss., 2010

Copyright © Peter Scholz
Copyright © 2011 Dodo Books Indian Ocean Ltd., member of the OmniScriptum S.R.L Publishing group

Für Kirstin

Abstract

This work focuses on the analysis and design of Inductive Power Transfer (IPT) antenna systems. Practical applications for IPT systems include a wireless powering of mobile devices in consumer electronics or Radio Frequency Identification (RFID) systems in logistics. The physical relevant properties of the antenna systems such as an accurate inductance computation or a precise modeling of skin and proximity effects are extracted by means of numerical techniques. At the same time, an equivalent network description based on the transformer concept is enabled by representing the antennas via reduced circuit models, which are obtained by specialized parameter fitting techniques.

The numerical simulations used in this thesis are based on the Partial Element Equivalent Circuit (PEEC) method. The PEEC method is especially appropriate for IPT antenna systems, because it allows efficient meshing techniques in case of long and thin conductors and provides a transformation of the electromagnetic coupling effects to the network domain. Furthermore, neglecting the retardation effects is traditionally fulfilled by the PEEC method when quasi-stationary assumptions of the Maxwell's equations are used. This is beneficial for IPT systems, since the simulation time is reduced while the errors are kept sufficiently small.

First, some fundamental concepts of electrodynamic effects are reviewed in this work. A new Lorenz-Quasi-Static (LQS) formulation is derived while its integration into well established techniques is shown. After presenting the fundamental concepts of IPT systems, the PEEC method is derived in a slightly modified way compared to the standard formulation in order to handle the different approximation techniques in a unified notation. Afterwards, the influence of parameter tolerances on the system behavior is analyzed by applying the adjoint sensitivity analysis to the PEEC method with a special focus on skin-effect problems. The presented system modeling approach is confirmed via measurements and Finite Element Method (FEM) simulations for a Printed Spiral Coil (PSC) system often used in RFID applications. By means of the optimized PEEC method, a remarkable speedup can be gained when compared with FEM simulations whereas the obtained errors typically do not exceed a few percent.

Contents

1 Introduction   1
   1.1 Motivation and Objective 1
   1.2 Inductive Power Transfer Applications 3
   1.3 Adequate Numerical Simulation Methods 4
   1.4 Outline of the Thesis 6

2 Classical Electrodynamics   9
   2.1 Maxwell's Equations 9
   2.2 Scalar Electric and Vector Magnetic Potentials 10
   2.3 Constitutive Equations 12
      2.3.1 Dielectric Materials 12
      2.3.2 Magnetic Materials 13
      2.3.3 Charge Density Inside Conductors 14
   2.4 Quasi-stationary Approximations 15
      2.4.1 Lorenz-Quasi-Static Formulation 18
      2.4.2 Magneto-Quasi-Static Formulation 21
   2.5 Green's Function Method 22
   2.6 Mixed Potential Integral Equation 23
   2.7 Poynting's Theorem 25
      2.7.1 Resistance 25
      2.7.2 Inductance 26
      2.7.3 Capacitance 29
      2.7.4 Impedance 31

3 Inductive Power Transmission   33
   3.1 Small Circular Loop Antenna 34
      3.1.1 Near- and Far-field Regions 37
      3.1.2 Resistive and Radiative Losses 38
      3.1.3 Inductance 40
   3.2 Design Constraints 41
      3.2.1 Frequency Range 41
      3.2.2 Quality Factor Definitions 42
      3.2.3 Rectangular Printed Spiral Coil 44

3.3	Equivalent Circuit Representation	45
	3.3.1 Air Coupled Transformer Concept	47
	3.3.2 Antenna Impedance Macromodeling	48
	3.3.3 System Design	56

4 Partial Element Equivalent Circuit Method — 65
- 4.1 Discretization — 66
- 4.2 Partial Network Elements — 71
- 4.3 Equivalent Circuit Representation — 74
  - 4.3.1 Nodal Based Analysis — 75
  - 4.3.2 Multi-Port Network — 78
- 4.4 Model Simplifications — 78
  - 4.4.1 Full-Wave and Quasi-Stationary — 79
  - 4.4.2 Magneto-Quasi-Static — 79
  - 4.4.3 Stationary Currents — 81
  - 4.4.4 2D Magneto-Quasi-Static — 81
- 4.5 Meshing Strategies — 82
  - 4.5.1 1D, 2D and 3D Meshes — 82
  - 4.5.2 Discretization of Conductor Bends — 83
  - 4.5.3 Panel Mesh of a Printed Spiral Coil — 84
  - 4.5.4 Mutual Inductance Computation of two Spiral Coils — 86
- 4.6 Modeling of Skin and Proximity Effects — 89
  - 4.6.1 State-of-the-Art Techniques — 89
  - 4.6.2 Subdivision of the Conductor's Cross Sections — 90
- 4.7 Modeling of Materials — 93
  - 4.7.1 Dielectric Materials — 93
  - 4.7.2 Magnetic Materials — 96
- 4.8 Acceleration Techniques — 96

5 Sensitivity Analysis — 99
- 5.1 Adjoint Based Method — 100
- 5.2 Inner-Layer Concept for Skin-Effect Sensitivities — 103

6 Simulation Results and Measurements — 107
- 6.1 Cylindrical Conductor — 107
  - 6.1.1 Solver Settings — 107
  - 6.1.2 Circular Cross Section and Infinite Length — 108
  - 6.1.3 Rectangular Cross Section and Infinite Length — 113
  - 6.1.4 Rectangular Cross Section and Finite Length — 117
  - 6.1.5 Sensitivity Analysis — 119
- 6.2 Printed Spiral Coil — 123

	6.2.1	Two Conductors Connected in Right Angle	124
	6.2.2	Rectangular Single-Turn Coil	127
	6.2.3	Rectangular Multi-Turn Antenna	131
6.3		Inductively Coupled Antenna System	143
	6.3.1	Setup of the Antenna System	143
	6.3.2	Mutual Inductance Computation	146
	6.3.3	Measurements of the Reader Antenna Input Impedance	147
	6.3.4	Measurements of the Data and Energy Transmission	147

7 Summary and Outlook — 153

A Partial Network Elements — 157
- A.1 Partial Inductances — 157
- A.2 Derivatives of the Partial Inductances — 162
- A.3 Partial Inductances in 2D — 164
- A.4 Partial Coefficients of Potential — 164
- A.5 Static Green's Function of a Two-layer Substrate — 166

B DC Analysis of a Rectangular Conductor Bend — 169

C Skin-Effect Discretization of a Rectangular Conductor — 177

Acronyms and Symbols — 181
- Acronyms — 181
- General Symbols and Conventions — 183
- Greek Letters — 183
- Roman Letters — 185

Bibliography — 189

Chapter 1
Introduction

In recent years, a growing interest in the wireless powering of mobile devices such as smartphones or laptops has emerged. Often, the battery charging process is the only remaining period in which the device has to be plugged to a charging platform via cables, since usually all data communication links have already been established wirelessly. Yet other applications aim to operate a receiving unit without a battery at all, which is often referred to as a passive device. These devices are remotely powered by one or more transmitting units either continuously or during pre-specified time slots. Reasons for operating devices in a passive way may be a low-cost producibility as in Radio Frequency Identification (RFID) applications or safety issues, for example if the device is implanted in living tissue. A further advantage of passive implanted devices is a generally much longer life cycle because the chemical processes inside the batteries are avoided, thus leading to a maintenance-free system. This feature is also important for passive systems in general and in particular for moving or rotating devices where a cabling is difficult if not impossible to realize. Last but not least, resulting improved product design capabilities if wires can be avoided should not be underestimated.

1.1 Motivation and Objective

The wireless transmission of electrical energy was first proposed by Tesla in the early 20th century [1]. In general, there exist two different mechanisms for transferring energy wirelessly. In the first case, classical electromagnetic waves are generated by a transmitting antenna and are picked up by a receiving antenna. The receiving unit converts the Radio Frequency (RF) energy of the waves to Direct Current (DC) energy which can be used to power the device, as can be seen in [2] for example. The main advantage of this approach is the ability to transport the energy over long distances especially if antennas with a high directivity are being used. However, this property requires a tracking ability for moving receivers and a line of sight between the transmitter and the receiver at any time of the power transmission. Moreover, since the power is radiated independently of the presence of a device, a communication link must be set up to properly control the power management. Due to the limited size of the antennas, applicable frequencies for

Chapter 1 Introduction

wireless power transfer via electromagnetic waves start at several MHz and may reach up to the THz regime if lasers are used to transfer the energy.

In the second variant, the power is not transferred via electromagnetic waves. Instead, use is made of the non-radiative electric and magnetic near-fields which are present in the vicinity of every antenna. In this case, the radiation of the powering antenna has to be minimized while maximizing either the magnetic or electric near-field. This is usually obtained by electrically small antennas at low and medium frequencies of up to several MHz. The main advantage of the near-field systems is the fact that, ideally, no power is transferred in the absence of a receiver. Instead, the electromagnetic energy is stored in the spatial region near the transmitter antenna until a receiver is brought in close proximity to the antenna. In this case, the energy can be picked up in order to power the device. Due to the near-field character of the system, there is an immediate back influence from the receiver to the transmitter, thus allowing the transceiver to detect the receiver.

The near-field coupling technique enables high efficiencies especially if resonant circuits in both transmitting and receiving units are being used. However, the drawback of the near-field wireless power transfer systems is the limitation to low- to medium-range applications with distances not significantly exceeding a few times the antenna dimensions, e.g. [3]. Furthermore, an inherent directivity of the near-fields of the antennas complicates the proper functionality for an arbitrary relative positioning and orientation of the devices.

Because of the low frequencies and the specific mounting forms of the antennas that are used in the near-field power transfer systems, the antennas are sometimes also referred to as coils or capacitors. Depending on whether the magnetic or electric energy dominates in the near-field region, a distinction between inductive and capacitive systems can be made. Compared to inductive systems, capacitive systems are less often encountered in practical applications because they react more sensitively to nearby everyday material with dielectric and metallic properties. Some investigations of transferring several watts over a distance of a few meters by means of an electric dominated near-field can be found, for instance, in [4].

In this work, special attention is paid towards wireless power systems which are based on inductive coupling. The reasons are, among others, the already mentioned high efficiency, almost no radiation and little interaction with environmental materials. In the following, inductive systems will be named Inductive Power Transfer (IPT) systems although some other terminologies such as *resonant energy transfer*, *resonant inductive coupling* or *electrodynamic inductive effect* have recently emerged.

IPT systems are based on the transformer concept which is known since Faraday's law in 1831 stating that a time-varying magnetic field caused by a primary current induces a voltage in a secondary current loop or coil. Hereby, a wireless power transfer between two different systems is enabled. The IPT systems which are addressed in this work differ from traditional transformers since no fixed coupling can be guaranteed and additionally no cores or at least no closed cores can be applied to guide the magnetic field. Moreover, IPT systems are generally implying a weak air coupling and are operated at higher frequencies compared to traditional transformers. These properties require a fast, accurate and efficient modeling and design technique for IPT antenna systems. The need for accurate design approaches is even more increased, since high efficiencies are aspired in IPT systems as aimed for all power transfer systems. This is equivalent to reducing the overall losses which are mainly evoked by eddy-current losses inside the conductors and substrates.

A detailed numerical analysis of the modeling of IPT antenna systems including different types of losses will be presented in later chapters. Prior to that, some IPT applications will be presented and a brief overview of numerical techniques being able to simulate the antenna systems with the aforementioned properties will be given.

1.2 Inductive Power Transfer Applications

In this section, a few applications are addressed in which IPT antenna systems are used or may be used in future. As mentioned above, the wireless powering and battery charging of mobile devices is of growing interest. In some experiments such as [5, 6], a successful transfer of several watts over distances of up to two meters has been demonstrated while reaching reasonable efficiencies.

A field of application with much less transferred power (microwatts to milliwatts) is given by the well established RFID technique which is a succeeding technology of the bar code systems used in the supply chain management. In order to uniquely identify various items, each object is tagged with a *Trans*mitter-Res*ponder* (Transponder) that can be identified by a reader unit via standard data communication techniques. In passive systems, the reader also powers the transponder either inductively or by radiation. Recently, some research has extended the traditional RFID principle from the pure identification of items to applications with sensors or displays, e. g. [7, 8]. Compared to the former identification applications, the energy demand of these so-called smart label applications is higher. If the system is operated inductively, a system design with a main focus on the IPT is required, consequently. Because the RFID technique with all its facets goes beyond the scope of this work, the reader is referred to [9] for more details. Nevertheless, past industrial projects such as [10] motivate the author to choose the antenna examples in the results chapter of this work based on the RFID technology.

Another field of research in which IPT systems have successfully been used since the early 1960ties [11] is given by biomedical applications and especially medical implants. In this case, the IPT is commonly known as *transcutaneous power transfer*. In [12], the inductive energy transmission has been investigated in order to provide energy to an auditory prosthesis while taking displacement tolerances into account. In [13], the power

Chapter 1 Introduction

transmission for an implanted biomedical device is enhanced in terms of optimizing the coupling coefficient of two spiral coils. A set of design rules is presented in [14] whereas a shape optimization of the coil system is focused on in [15].

A third application field concerns coreless planar Printed Circuit Board (PCB) transformers, (cf. [14, 16]), which aim to miniaturize transformers for microelectronic applications. The generally higher frequencies and the lack of the field-guiding cores make such a system design comparable to the previously mentioned applications although a fixed coupling can be ensured. However, additional difficulties may occur for densely packaged miniaturized applications. The coil design in integrated circuits such as needed for integrated Voltage Controlled Oscillators (VCOs), e. g. [17] is also addressed in this work because similar challenges including high quality factors may arise. In contrast to the former applications, only a single coil is typically used in the VCO design.

Last but not least, the IPT approach can also be used to power moving vehicles such as buses or special transporting systems. This offers new kinds of applications in which batteries of electrically powered automobiles are charged inductively. Due to the high power demand of these systems and the resulting design challenges, low frequencies in the kHz range are traditionally preferred.

1.3 Adequate Numerical Simulation Methods

In order to design and optimize the antennas of IPT systems, appropriate design approaches are required. The applicability of analytical expressions is analyzed first, since these provide by far the fastest way to obtain information about the system behavior. A multitude of approximative expressions especially for computing the self- and mutual inductances of different coil geometries can be found, e. g., in [9, 18, 19, 20, 21]. The equations in the references are either derived by empirical studies or by using different approximation techniques and are mostly concerning simple geometries and orientations. Consequently, the applicability is restricted to rough estimations for initial system design purposes. In addition, no closed-form mutual inductance extraction technique is known for spiral coils with arbitrary reciprocal orientation.

The application of analytical methods becomes even less feasible if frequency-dependent eddy-current losses which are causing skin and proximity effects cannot be neglected. This is especially the case when the efficiency of the overall system behavior has to be maximized. Furthermore, the capacitive couplings of the conductors are traditionally not included in the analytical expressions.

In order to provide a design alternative to a development approach by means of measurements, numerical antenna design tools which are based on a discrete formulation of the Maxwell's equations are suggested. An adequate numerical method should be able to compute the near-field coupling of arbitrary 3D antenna structures with an accurate loss determination and a fast simulation time allowing for spatial parameter sweeps.

1.3 Adequate Numerical Simulation Methods

Preferably, a network description based on the transformer concept should be obtained and a sensitivity analysis could allow the examination of parameter tolerances on the system. If quasi-stationary assumptions of the Maxwell's equations can simplify the calculations, they are preferred over full wave analysis because radiation effects can be neglected in most cases.

A comparison of different numerical methods to solve Electromagnetic (EM) problems can be found in [22, 23] for instance. From a technical point of view there exist two inherently different approaches for simulating EM problems. In particular, a distinction can be made into methods either discretizing the full volume or just the individual materials located in the considered calculation domain.

The former methods are mainly based on the differential form of the Maxwell's equations and discretize the computational domain into elementary volume cells in which the underlying equations are fulfilled in a local sense. This leads to sparse matrix formulations, because the cells are only coupled with their neighbors. Since each elementary cell may have different material properties, these numerical methods are very flexible and thus being suitable for a wide range of applications. Two exemplary methods of this class of approaches are the Finite Element Method (FEM), (cf. [24]) and the Finite Integration Technique (FIT) [25].

Contrary to this, the latter methods are based on integral equations which are deduced from the Maxwell's equations. By using the Green's function method, the EM problem is solved by a superposition of elementary solutions which are automatically fulfilling open boundary conditions. In the general context of the Method of Moments (MoM) [26], a typical integral equation based method in electrodynamics is the Boundary Element Method (BEM) [27], in which only the surfaces of the homogeneous materials are discretized.

The properties of integral equation based methods are often advantageous for open problems with large regions of free space as is often the case in antenna or scattering problems. For these problems, a much smaller system matrix compared to the FEM or FIT is obtained. Thus, the simulation time can be substantially reduced although the matrices are dense due to the coupling of all elements with each other. A further reduction of the simulation time can be achieved by using specialized matrix compression techniques such as the Fast Multipole Method (FMM). The main drawback of integral equation methods is the difficulty to handle inhomogeneous, nonlinear or anisotropic material distributions, thus lowering the generality of these methods.

In the case of interconnection structures, a specific realization of the MoM is the Partial Element Equivalent Circuit (PEEC) method [28] which is based on the Mixed Potential Integral Equation (MPIE). The PEEC method naturally transforms the electromagnetic field problem into an equivalent *RLC* network representation by using piecewise constant basis and testing functions. The obtained partial circuit elements

Chapter 1 Introduction

are connected according to Kirchhoff's current and voltage laws and can be analyzed via circuit solving packages such as SPICE.

The PEEC method is particularly suitable for the simulation of IPT antenna systems for several reasons. First of all, real world IPT antenna systems are generally located in non-bounded space and the proportion of free space compared to the occurring materials especially the conductors is generally very high. This makes integral equation based methods beneficial as mentioned before. Second, because of the typically utilized medium-range frequencies in IPT systems, quasi-stationary assumptions are favorable. These assumptions are traditionally fulfilled by the PEEC method. Third, eventually occurring eddy-current losses can be accounted for by a volume discretization of the conductors. The network character of the PEEC method is an additional benefit because the external circuitry can be considered in a natural way. Two further inherent properties of the PEEC method allow for speeding up the simulations by reducing unknowns. This includes a pre-limiting of the elements to the estimated current direction on the one hand and a building of the cells with high aspect ratios on the other hand. More details about the PEEC method and the specialized mesh settings will be presented in later chapters.

Besides the mentioned advantages, some limitations of the PEEC method should not be concealed. If the spatial domain is filled with large objects of conducting, dielectric or magnetic material in which a 3D discretization must be set up, the system size increases dramatically and some of the aforementioned advantages over sparse matrix methods are lost. Nevertheless, the PEEC method will be used throughout this work whereas the numerical results are compared with FEM results as well as measurements.

Although a number of commercial and non-commercial PEEC tools such as CST PCB STUDIO™ [29] or FastHenry [30] are available, a specialized PEEC solver has been developed in the course of this work in order to enable flexibility in terms of mesh generation and integral evaluation. Developing a specialized code allows, in particular, to combine different mesh settings and quasi-stationary solver setups. This is advantageous for extracting reduced network models which can be used to characterize the IPT antenna system. In addition, a sensitivity analysis has been implemented in order to quantify the influence of parameter tolerances on the system behavior.

1.4 Outline of the Thesis

This thesis is structured as follows. In chapter 2, the fundamental concepts of classical electrodynamics which are needed for the following chapters are briefly presented. The chapter focuses on the quasi-stationary field approximations, since these assumptions simplify the complexity of the underlying equations and provide good approximations for IPT systems. In order to gain a better insight into the inductive and capacitive effects from an energetic point of view and to consequently use both effects in a common analysis, a new Lorenz-Quasi-Static (LQS) approach is derived which closes the gap between full-wave analysis and the commonly known Electro-Quasi-Static (EQS) and Magneto-Quasi-Static (MQS) formulations.

In chapter 3, the necessary concepts for IPT antenna systems are presented. Besides the physical relevant fundamentals which are illustrated for a circular loop antenna, important design factors such as different quality factor definitions are presented. Afterwards, the system design is analyzed in terms of an equivalent network description with focus on the extraction of reduced antenna models and the optimization of the overall system behavior in terms of efficiency.

In chapter 4, the PEEC method is derived in frequency domain using a slightly different notation compared to the standard work [28] in order to account for the different quasi-stationary assumptions considered in this work. Throughout the derivation of the method, the particularities concerning the modeling of IPT systems are highlighted. Especially the mesh settings that differ for the inductive and capacitive meshes as well as for self-impedance and mutual inductance computations are discussed.

In chapter 5, the adjoint sensitivity analysis is reviewed and the applicability to the PEEC method is shown. A technique for optimizing the method for skin-effect problems is proposed.

In chapter 6, numerical results of the PEEC method are presented for an individual conductor, a single coil as well as for an IPT coil system consisting of an RFID reader single-turn and a transponder multi-turn coil. In order to verify and validate the results, comparisons with exact analytical expressions, numerical FEM simulations as well as measurements are presented. It will be demonstrated that for the coil design, PEEC simulations can be performed within seconds to minutes whereas comparable FEM simulations may last hours to days. Furthermore, a fast mutual inductance computation based on a coarse PEEC mesh allows for precisely forecasting the powering range of arbitrary 3D IPT antenna arrangements within a few milliseconds.

A summary recapitulates the main results of this work as well as it provides a short outlook to further studies.

Chapter 1 Introduction

Chapter 2

Classical Electrodynamics

In this chapter, a few aspects of the classical electrodynamic field theory are reviewed in order to introduce the fundamental concepts which are needed for the subsequent chapters. Starting with the Maxwell's equations, the electric scalar potential as well as the magnetic vector potential are introduced and the resulting wave equations are derived. Then, the quasi-stationary approximations are discussed. In this context, a new LQS formulation which is based on the potentials is derived and its compatibility with standard formulations is shown. Solutions to the full-wave as well as quasi-stationary formulations are given by means of the Green's function method and an integral equation formulation required by the PEEC method is presented. The last section concentrates on the definitions of the resistance, inductance, capacitance and impedance which are of importance for the network description employed in later chapters.

2.1 Maxwell's Equations

Electromagnetic (EM) field problems are described by Maxwell's equations which can be expressed in differential form in time and frequency domain as

Time domain: *Frequency domain:*

$$\operatorname{curl} \vec{E} = -\frac{\partial \vec{B}}{\partial t} \qquad \operatorname{curl} \underline{\vec{E}} = -j\omega \underline{\vec{B}} \qquad (2.1\mathrm{a})$$

$$\operatorname{curl} \vec{H} = \frac{\partial \vec{D}}{\partial t} + \vec{J} \qquad \operatorname{curl} \underline{\vec{H}} = j\omega \underline{\vec{D}} + \underline{\vec{J}} \qquad (2.1\mathrm{b})$$

$$\operatorname{div} \vec{D} = \varrho \qquad \operatorname{div} \underline{\vec{D}} = \underline{\varrho} \qquad (2.1\mathrm{c})$$

$$\operatorname{div} \vec{B} = 0 \qquad \operatorname{div} \underline{\vec{B}} = 0. \qquad (2.1\mathrm{d})$$

In frequency domain, the time derivatives $\partial/\partial t$ are replaced by the factor $j\omega$ with j being the imaginary unit and ω the angular frequency, respectively. In order to distinguish

Chapter 2 Classical Electrodynamics

between time- and frequency-domain formulations, the complex amplitudes are denoted by underlined symbols. The following derivations will be performed in time domain in order to preserve generality. The frequency domain will be used whenever it is convenient for the analysis. For better readability, the explicit dependencies on space $\vec{r}$, time t and frequency ω are omitted except it is stated otherwise.

The vector fields appearing in (2.1) are the electric field strength $\vec{E}(\vec{r},t)$, the electric flux density $\vec{D}(\vec{r},t)$, the magnetic field strength $\vec{H}(\vec{r},t)$ and the magnetic flux density $\vec{B}(\vec{r},t)$, respectively. The sources are specified by the electric current density $\vec{J}(\vec{r},t)$ and the electric charge density $\varrho(\vec{r},t)$. The electric and magnetic field strengths and fluxes are linked by the constitutive equations

$$\vec{D} = \varepsilon_0 \vec{E} + \vec{P}, \tag{2.2a}$$

$$\vec{H} = \frac{1}{\mu_0}\vec{B} - \vec{M}. \tag{2.2b}$$

The material constants ε_0 and μ_0 are the permittivity and permeability of the free space while $\vec{P}(\vec{r},t,\vec{E})$ denotes the polarization and $\vec{M}(\vec{r},t,\vec{B})$ the magnetization of the medium, respectively. These vector fields describe the macroscopic behavior of the physical effects inside the materials, generally depending on the electric field strength or magnetic flux density. In the following, the polarization and magnetization are treated as electromagnetic source fields in addition to the usual currents and charges. When solving a specific type of problem, the dependence of these quantities on the fields must be regarded in the resulting set of equations.

2.2 Scalar Electric and Vector Magnetic Potentials

Following the usual derivation in standard text books (s. [31] for example), it is convenient to express the electric field strength and the magnetic flux density by a scalar electric potential $\Phi(\vec{r},t)$ and a magnetic vector potential $\vec{A}(\vec{r},t)$ as

$$\vec{B} = \operatorname{curl} \vec{A}, \tag{2.3a}$$

$$\vec{E} = -\operatorname{grad} \Phi - \frac{\partial \vec{A}}{\partial t}. \tag{2.3b}$$

By the introduction of the potentials, the two Maxwell's equations (2.1a) and (2.1d) are satisfied implicitly because of the vector identities $\operatorname{div} \operatorname{curl} \vec{F} = 0$ and $\operatorname{curl} \operatorname{grad} \vec{F} = 0$ being valid for any vector field $\vec{F}$. The potentials allow for converting the original system of coupled partial differential equations into a smaller one with higher order that still satisfies Maxwell's equations.

By using the potentials it is possible to set up two coupled differential equations by successively substituting (2.3b) and (2.2a) into (2.1c) for the scalar and (2.3) as well as

2.2 Scalar Electric and Vector Magnetic Potentials

(2.2b) into (2.1b) for the vector potential which leads to

$$\Delta \Phi + \frac{\partial}{\partial t} \operatorname{div} \vec{A} = -\frac{1}{\varepsilon_0} \varrho_{\text{tot}}, \qquad (2.4a)$$

$$\Delta \vec{A} - \frac{1}{c_0^2} \frac{\partial^2 \vec{A}}{\partial t^2} - \operatorname{grad}\left(\operatorname{div} \vec{A} + \frac{1}{c_0^2} \frac{\partial \Phi}{\partial t}\right) = -\mu_0 \vec{J}_{\text{tot}}, \qquad (2.4b)$$

with $c_0 = 1/\sqrt{\varepsilon_0 \mu_0}$ being the speed of light in vacuum. In (2.4), the following abbreviations have been introduced

$$\varrho_{\text{tot}} = \varrho + \varrho^{\text{P}}, \qquad \vec{J}_{\text{tot}} = \vec{J} + \vec{J}^{\text{P}} + \vec{J}^{\text{M}}, \qquad (2.5a)$$

where the polarization charge density $\varrho^{\text{P}}(\vec{r}, t, \vec{E})$, the polarization current density $\vec{J}^{\text{P}}(\vec{r}, t, \vec{E})$ and the magnetization current density $\vec{J}^{\text{M}}(\vec{r}, t, \vec{B})$ are defined as

$$\varrho^{\text{P}} = -\operatorname{div} \vec{P}, \qquad \text{Polarization charge density}, \qquad (2.5b)$$

$$\vec{J}^{\text{P}} = \frac{\partial \vec{P}}{\partial t}, \qquad \text{Polarization current density}, \qquad (2.5c)$$

$$\vec{J}^{\text{M}} = \operatorname{curl} \vec{M}, \qquad \text{Magnetization current density}. \qquad (2.5d)$$

In (2.5a), the subscripts "tot" indicate the total charge and current densities induced by polarization and magnetization effects as well as those impressed by external sources.

The choice of the potentials in (2.3) is not unique. In particular, the divergence of $\vec{A}$ can be chosen arbitrarily. A commonly gauging (fixing) for the vector potential $\vec{A}$ is the so-called Lorenz gauge

$$\operatorname{div} \vec{A} = -\frac{1}{c_0^2} \frac{\partial \Phi}{\partial t}. \qquad (2.6)$$

Using (2.6), equations (2.4) are decoupled leading to the symmetric form

$$\Delta \Phi - \frac{1}{c_0^2} \frac{\partial^2 \Phi}{\partial t^2} = -\frac{1}{\varepsilon_0} \varrho_{\text{tot}}, \qquad (2.7a)$$

$$\Delta \vec{A} - \frac{1}{c_0^2} \frac{\partial^2 \vec{A}}{\partial t^2} = -\mu_0 \vec{J}_{\text{tot}}. \qquad (2.7b)$$

Equations (2.7) describe a system of inhomogeneous wave equations which are coupled via the Lorenz gauge in (2.6) or, alternatively, via the continuity equation. This can be verified by applying the Laplace operator to (2.6), using the vector identity $\Delta \operatorname{div} \vec{A} = \operatorname{div} \Delta \vec{A}$, inserting (2.7), rearranging terms and inserting (2.6) again which results in the continuity equation

$$\operatorname{div} \vec{J}_{\text{tot}} + \frac{\partial}{\partial t} \varrho_{\text{tot}} = 0. \qquad (2.8a)$$

By using the definitions of (2.5), the continuity equation can be extended to the particular currents and charges as

$$\operatorname{div} \vec{J} + \frac{\partial \varrho}{\partial t} = 0, \qquad (2.8\text{b})$$

$$\operatorname{div} \vec{J}^{\mathrm{P}} + \frac{\partial}{\partial t} \varrho^{\mathrm{P}} = 0, \qquad (2.8\text{c})$$

$$\operatorname{div} \vec{J}^{\mathrm{M}} = 0. \qquad (2.8\text{d})$$

Using the scalar and vector potentials, the electromagnetic field problem is completely described by the wave equations (2.7) and the continuity equation (2.8a). The electric and magnetic fields may be computed from the potentials by means of (2.3) and (2.2). A general solution of the wave equations via the Green's function method will be presented in section 2.5. It should be mentioned that different gauges for the vector potential $\vec{A}$ may result in different solutions of the potentials but do not affect the solutions for the electric and magnetic fields.

2.3 Constitutive Equations

In the last section, the polarization and the magnetization $\vec{P}$ and $\vec{M}$ have been treated as source terms although they are, generally, time-variant, frequency dependent, non-linear as well as non-isotropic functions of $\vec{E}$ and $\vec{B}$, respectively. Because a modeling of such a general material behavior is typically difficult to handle, the following considerations are restricted to materials with linear, time-invariant and isotropic behavior. These materials are focused on in the following subsections by deriving more detailed expressions.

2.3.1 Dielectric Materials

For linear, time-invariant, non-dispersive and isotropic media, the dependence of the polarization vector $\vec{P}$ on the electric field strength can be expressed by a single scalar quantity which is either its relative permittivity $\varepsilon_{\mathrm{r}}(\vec{r})$ or total permittivity $\varepsilon(\vec{r}) = \varepsilon_0 \varepsilon_{\mathrm{r}}(\vec{r})$ according to

$$\vec{P} = (\varepsilon - \varepsilon_0)\vec{E} = \varepsilon_0(\varepsilon_{\mathrm{r}} - 1)\vec{E}. \qquad (2.9)$$

When this definition is substituted in (2.2a), the well-known expression for the electric flux density is obtained:

$$\vec{D} = \varepsilon_0 \vec{E} + \vec{P} = \varepsilon_0 \varepsilon_{\mathrm{r}} \vec{E}. \qquad (2.10)$$

2.3 Constitutive Equations

Surface Polarization Charge It can be shown that for piecewise homogeneous dielectrics, polarization charges can exist only on the surfaces of the materials.[1] This is an important matter of fact because the complexity of the EM problem can be reduced by restricting the charges on boundaries. Surface polarization charges are often called bounded charges in contrast to the charges inside of the conductors which are also known as free charges.

For proving the above mentioned statement, the polarization charge density (2.5b) is expressed as a function of $\vec{E}$ while inserting (2.9) as

$$\varrho^{\mathrm{P}} = -\operatorname{div}\left(\varepsilon_0 \left[\varepsilon_{\mathrm{r}}(\vec{r}) - 1\right] \vec{E}\right). \tag{2.11a}$$

By using the vector relation $\operatorname{div}(\Theta \vec{F}) = \Theta \operatorname{div} \vec{F} + \vec{F} \cdot \operatorname{grad} \Theta$ for two arbitrary scalar- and vector fields Θ and $\vec{F}$, the above equation can be rearranged while substituting $\operatorname{div}(\varepsilon_0 \vec{E}) = \varrho + \varrho^{\mathrm{P}}$ by (2.2a), (2.5b) and (2.1c) leading to

$$\varrho^{\mathrm{P}} = \frac{1 - \varepsilon_{\mathrm{r}}}{\varepsilon_{\mathrm{r}}} \varrho - \frac{\varepsilon_0}{\varepsilon_{\mathrm{r}}} \vec{E} \cdot \operatorname{grad} \varepsilon_{\mathrm{r}}(\vec{r}). \tag{2.11b}$$

For every $\vec{r}$ located inside a homogeneous dielectric material which can be characterized by a constant ε_{r}, both terms in the right hand side of (2.11b) vanish. The first one because the free charge density ϱ inside the dielectric material is zero and the second one because ε_{r} is constant. Thus, the polarization charge can only be located at the surfaces of the dielectric regions where the discontinuity of the dielectric material has a non-vanishing gradient in (2.11b).

Complex Permittivity Losses inside a dielectric material, e.g., due to a nonzero conductivity, can be accounted for in the frequency domain by combining the current density and the displacement current density of (2.1b) to a complex permittivity $\underline{\varepsilon} = \varepsilon + \kappa/(j\omega)$. A commonly used formulation for the relative complex permittivity is

$$\underline{\varepsilon}_{\mathrm{r}} = \varepsilon_{\mathrm{r}} \left(1 - j \tan \delta\right), \tag{2.12}$$

in which $\tan \delta$ is the loss tangent of the material. In general, ε_{r} as well as $\tan \delta$ are frequency dependent.

2.3.2 Magnetic Materials

The derivations of the last section can be similarly applied to the case of magnetic materials. Equivalently to the above, for linear, time-invariant, non-dispersive and isotropic media, the magnetization can be expressed by a scalar permeability factor

$$\vec{M} = (\mu_{\mathrm{r}} - 1) \vec{H}, \tag{2.13}$$

---
[1] Materials that are specifically doped with a volume charge density are not covered by this discussion.

which is given either by its relative permeability $\mu_r(\vec{r})$ or by the total permeability $\mu(\vec{r}) = \mu_0 \mu_r(\vec{r})$. Inserting (2.13) into (2.2b) yields the constitutive relation for magnetic fields:

$$\vec{B} = \mu_0 \mu_r \vec{H}. \tag{2.14}$$

Surface Magnetization Current Assuming that neither a conducting current nor a displacement current density exist inside a piecewise homogeneous magnetic material, the magnetization current is limited to the boundary surface of the material. In order to prove this statement, the magnetization current density is written as a function of the magnetic field intensity as well as the relative permeability by inserting (2.13) into (2.5d) and using the vector relation $\operatorname{curl}(\Theta \vec{F}) = \Theta \operatorname{curl} \vec{F} - \vec{F} \times \operatorname{grad} \Theta$ which results in

$$\vec{J}^M = [\mu_r(\vec{r}) - 1] \operatorname{curl} \vec{H} - \vec{H} \times \operatorname{grad} \mu_r(\vec{r}). \tag{2.15}$$

It can be seen that for constant μ_r, the gradient in the last term of (2.15) is identically zero. The first term vanishes in the trivial case $\mu_r = 1$ or if $\operatorname{curl} \vec{H} = 0$. In (2.1b), the curl of $\vec{H}$ is composed of two parts; the conducting current density $\vec{J}$ and the displacement current density $\partial \vec{D}/\partial t$. If both of them vanish, e.g. for a magnetic material with zero conductivity and for static fields, only a surface magnetization current is present. In all other cases, however, the magnetization current density does not vanish inside the magnetic material.

2.3.3 Charge Density Inside Conductors

In this section, it will be shown that the charge density inside homogeneous conductors with a sufficiently high conductivity can be assumed to be zero for almost any practical application. This knowledge can be used in a numerical method to a priori limit the unknown charges to the surfaces of the conductors. The derivation is similar to the preceding sections although the form of the equation is slightly different. The current density in conductors is given by Ohm's law

$$\vec{J} = \kappa \vec{E}, \tag{2.16}$$

where $\kappa(\vec{r})$ is the electric conductivity of the material. The continuity equation (2.8b), using (2.16), (2.10) and (2.1c) can be written as

$$\operatorname{div} \vec{J} = \operatorname{div}\left(\frac{\kappa(\vec{r})}{\varepsilon(\vec{r})} \vec{D}\right) = \frac{\kappa(\vec{r})}{\varepsilon(\vec{r})} \varrho + \vec{D} \cdot \operatorname{grad}\left(\frac{\kappa(\vec{r})}{\varepsilon(\vec{r})}\right) \tag{2.17a}$$

$$= -\frac{\partial}{\partial t} \varrho. \tag{2.17b}$$

For any point $\vec{r}$ inside a homogeneous conductor with constant κ and ε, the gradient in the last term of (2.17a) vanishes. Furthermore, the charge density ϱ in (2.17) can be shown to decrease exponentially, since

$$\frac{\partial \varrho}{\partial t} + \frac{\kappa}{\varepsilon}\varrho = 0, \quad \Rightarrow \quad \varrho(\vec{r},t) = \varrho(\vec{r},t_0)\, e^{-(\kappa/\varepsilon)\,t}. \tag{2.18}$$

From this equation it is obvious that a given initial charge distribution $\varrho(\vec{r},t_0)$ at $t=t_0$ will decay exponentially with the relaxation time $\tau_e = \varepsilon/\kappa$. As an example, a conductor made of copper has a relaxation time of about $\tau_e = 1.5\,10^{-19}$ s. If this time is much smaller than the rise time of the investigated signal, the volume charge density can be neglected. On the surface of the conductors, the gradient in (2.17a) differs from zero and contributes to a surface charge density $\sigma(\vec{r},t)$.

2.4 Quasi-stationary Approximations

Physics based approximations to the full set of Maxwell's equations are widely used for static or low frequency problems in which the wave character does not significantly influence the overall system behavior. Such approximations are beneficial because the complexity of the underlying set of equations can be reduced. The specific approximation type is selected by either engineering expertise or some rules of thumb which predict its range of validity. The common way is to validate the result obtained by using the approximate formulation with a full-wave reference solution or, alternatively, with measurements.

The approach used in this work is different, as two different low frequency approximations are applied simultaneously. This allows for a deduction of the physical relevant properties of the system. In the following, two low frequency approximations are systematically derived. An algorithm to combine the two formulations will be discussed in section 3.3.2.

Approximations of Maxwell's equations for slowly time varying fields are referred to as quasi-stationary assumptions. These formulations share the property of neglecting radiation and retardation effects in the underlying Maxwell's equations. This is motivated by the fact that radiation and retardation are physically not relevant at low- and medium frequencies. Thus, neglecting them does not affect the accuracy of the solution while typically leading to a simplified set of field equations compared to the full set of Maxwell's equations. The traditional way to obtain such low frequency approximations is to either neglect the magnetic induction term $\partial \vec{B}/\partial t$ in (2.1a) or the displacement current density $\partial \vec{D}/\partial t$ in (2.1b) leading to the Electro-Quasi-Static (EQS) and Magneto-Quasi-Static (MQS) formulations, respectively, e. g. [32, 33]. Both approximations do not contain radiation effects; EQS is used for applications with dominating capacitive effects while MQS describes applications with dominating inductive effects.

Chapter 2 Classical Electrodynamics

It is, however, sometimes necessary to consider both, inductive *and* capacitive effects, e. g. when analyzing the Self-Resonant Frequency (SRF) of an inductor. This behavior can be captured by neither EQS nor MQS because in the first case the magnetic energy is non-existent while in the latter case the electric energy, respectively [34]. In order to overcome this situation, the common alternative is to apply a full-wave approach which could be unnecessarily complicated since radiation is still negligible for electrically small antennas.

This example motivates to search for a further refinement of the quasi-stationary approximations. The contributions [34, 35, 36, 37, 38] discuss the topic more detailed and share the same key idea which is basically not to neglect the whole displacement current of (2.1b) but only a fraction of it. By doing so, a more accurate approximation than EQS and MQS is obtained which can capture capacitive *and* inductive behavior simultaneously. In fact, using an appropriate decomposition of the electric field, a whole hierarchy of low frequency approximations can be thought of. For this purpose, a series representation of the fields is used in [34, 35, 39], while the formulation presented in [36, 38] decomposes the electric field strength $\vec{E}$ and the electric flux density $\vec{D}$ into two parts each, the irrotational (curl-free) and solenoidal (divergence-free) one. Due to the Helmholtz's theorem, this decomposition is unique when assuming that the fields are sufficiently smooth and rapidly decaying at infinity. Followed by that, the approximation to the full set of the Maxwell's equations is to neglect the solenoidal part of the displacement current in (2.1b).[2]

Due to the uniqueness of the field decomposition, the methodology works independently of the definition of the potentials from section 2.2. Whenever the potentials of (2.3) are used to describe the EM system, the above decomposition scheme naturally uses the Coulomb gauge div $\vec{A} = 0$. This is because the Coulomb gauge identically maps the electric field of (2.3b) into its irrotational and solenoidal parts because of curl grad $\Phi = 0$. In this gauge, the free-space potential equations (2.7) become in the approximation [36, 37]

$$\Delta \Phi = -\frac{\varrho}{\varepsilon_0}, \tag{2.19a}$$

$$\Delta \vec{A} = -\mu_0 \vec{J} + \frac{1}{c_0} \operatorname{grad} \frac{\partial \Phi}{\partial t}. \tag{2.19b}$$

As desired, these equations do not show radiation effects in contrast to the wave equations (2.7) due to the missing terms with the double differentiation in time. However, the drawback of the above formulation is the fact that the last term of the right hand

---

[2] In the context of interacting charged particles in free space, this methodology is also known as Darwin formulation introduced in 1920 [40], (s. also [37] and the references therein). In [38] it is stated that the incorporation of the irrotational part of the displacement current only has first been proposed by Clausius in between 1875 and 1877.

2.4 Quasi-stationary Approximations

side of (2.19b) can be interpreted as an additional current part[3] which is generally not restricted to the conducting regions in contrast to the conduction current density $\vec{J}$.

This term complicates the EM problem formulation especially for numerical methods based on integral equations which use currents and charges as the unknowns. This is because a discretization of the full spatial domain instead of the material regions only is required.[4] This is typically bypassed by letting $c_0 \to \infty$ in (2.19b) which leads to the standard magneto-static and MQS expression for the vector potential.[5] It should be noted that the same result could have been reached by letting $c_0 \to \infty$ in (2.7). However, implying an infinite speed of light formally requires either ε_0 or μ_0 to be equal to zero because of $c_0 = 1/\sqrt{\varepsilon_0\mu_0}$ [34].

In order to obtain a better understanding of the underlying system of equations, it is aimed in this section to find a new approximative formulation of the Maxwell's equations that should maintain the property of neglecting the radiation and in addition should be consistent with a new basic system of equations. The formulation should be able to capture capacitive and inductive effects according to the above mentioned formulation but should *not* include the additional current part in the right hand side of (2.19b) which is generally nonzero in the whole spatial domain. It will turn out that a solution can be found via a similar decomposition scheme of $\vec{E}$ and $\vec{D}$ as in the Helmholtz decomposition. The difference of the approach presented here is the fact that the decomposition is no longer defined by the irrotational and solenoidal parts but instead is only determined by the potentials Φ and $\vec{A}$. This of course limits the applicability of the new formulation to mathematical techniques which are based on the potentials.

Explicit use will be made of the Lorenz gauge (2.6) which releases the property of the originally divergence-free part of the electric field in the above formulation based on the Helmholtz decomposition. It should be mentioned that in contrast to the full set of the Maxwell's equations which are invariant to a gauge transformation, this is generally no longer valid for the quasi-stationary approximations when the decomposition of the electric field depends on the potentials themselves. Due to the specific choice of the decomposition by means of the Lorenz gauge, the new system necessitates a further approximation w.r.t. the Coulomb gauge which generally reduces the range of applicability.

The new formulation will be named Lorenz-Quasi-Static (LQS) in order to account for the difference to the aforementioned formulation in which the Coulomb gauge is applied. In the following subsections, the LQS formulation will be derived. Followed by that, the standard MQS approximation will shortly be reviewed. Both formulations result in two different basic sets of equations and are both needed for the following chapters which will focus on IPT systems and on the PEEC method. It will be seen that the

---

[3]More specifically, it is a part of the displacement current density.
[4]Assuming that no Green's function accounting for the $\operatorname{grad} \partial\Phi/\partial t$ term is known to solve (2.19).
[5]The system (2.19) still differs from the static or MQS cases because of the continuity equation (2.8b).

two models distinguish each other only in minor parts of the PEEC solver and can therefore be implemented with small code changes only. The main advantage of using the two formulations is the fact that by simulating both systems, physical relevant parameters that are needed for the macromodels such as inductances and capacitances can be extracted by comparing the results. Application of this approach enables an easy and physically motivated parameter extraction for reduced circuit models.

2.4.1 Lorenz-Quasi-Static Formulation

This section systematically derives the decomposition of the electric field quantities into quasi-static and induced parts. The aim is to identify the retardation parts of (2.7) with double differentiation in time that are responsible for the radiation in the Maxwell's equations. For this reason, these terms are brought to the right hand side of the wave equations

$$\Delta \Phi = -\frac{1}{\varepsilon_0}\varrho_{\text{tot}} - \text{div}\left(\frac{\partial \vec{A}}{\partial t}\right), \tag{2.20a}$$

$$\Delta \vec{A} = -\mu_0 \vec{J}_{\text{tot}} + \frac{1}{c_0^2}\frac{\partial}{\partial t}\left(\frac{\partial \vec{A}}{\partial t}\right), \tag{2.20b}$$

where in (2.20a), the Lorenz gauge (2.6) has been substituted. It is seen that if $\partial \vec{A}/\partial t$ would be zero in both equations, the radiation would have vanished. In order to find an approximated version of the Maxwell's equations that does not include the $\partial \vec{A}/\partial t$ term in (2.20), it is necessary to decompose the electric field strength from (2.3b) as

$$\vec{E} = -\operatorname{grad}\Phi - \frac{\partial \vec{A}}{\partial t} = \vec{E}_0 + \vec{E}_i, \tag{2.21a}$$

with introducing

$$\vec{E}_0 = -\operatorname{grad}\Phi, \qquad \text{Quasi-static electric field strength,} \tag{2.21b}$$

$$\vec{E}_i = -\frac{\partial \vec{A}}{\partial t}, \qquad \text{Induced electric field strength.} \tag{2.21c}$$

Because the decomposition scheme via $\vec{E}_0$ and $\vec{E}_i$ will be used in the new LQS formulation, it is obvious that this scheme cannot be used without the introduction of the potentials. When comparing (2.20) and (2.21), $\vec{E}_i$ is found to be responsible for the radiation terms in (2.20). Due to the assumed linearity of the materials, a subsequent

2.4 Quasi-stationary Approximations

decomposition of $\vec{P}$, $\vec{D}$, ϱ and ϱ^{P} is obtained in a straightforward manner as

$$\vec{D} = \underbrace{\underbrace{\varepsilon_0 \vec{E}_0 + \overbrace{(\varepsilon - \varepsilon_0)\vec{E}_0}^{\vec{P}_0}}_{\vec{D}_0} + \underbrace{\varepsilon_0 \vec{E}_\mathrm{i} + \overbrace{(\varepsilon - \varepsilon_0)\vec{E}_\mathrm{i}}^{\vec{P}_\mathrm{i}}}_{\vec{D}_\mathrm{i}}}, \qquad (2.22\mathrm{a})$$

$$\varrho = \underbrace{\mathrm{div}\,\varepsilon_0 \vec{E}_0 + \overbrace{\mathrm{div}\,\vec{P}_0}^{-\varrho_0^{\mathrm{P}}}}_{\varrho_0} + \underbrace{\mathrm{div}\,\varepsilon_0 \vec{E}_\mathrm{i} + \overbrace{\mathrm{div}\,\vec{P}_\mathrm{i}}^{-\varrho_\mathrm{i}^{\mathrm{P}}}}_{\varrho_\mathrm{i}}. \qquad (2.22\mathrm{b})$$

Equations (2.22) show that both $\vec{D}$ and ϱ are composed of multiple parts each in general. If the whole space is composed of homogeneous material ε_0, no polarization exists and the above equations simplify.

With the decomposition scheme (2.21) and (2.22) it is possible to write the new Lorenz-Quasi-Static (LQS) set of equations that approximate Maxwell's equations (2.1) as

$$\mathrm{curl}\,\vec{E} = -\frac{\partial \vec{B}}{\partial t} \qquad (2.23\mathrm{a})$$

$$\mathrm{curl}\,\vec{H} = \frac{\partial \vec{D}_0}{\partial t} + \vec{J} \qquad (2.23\mathrm{b})$$

$$\mathrm{div}\,\vec{D}_0 = \varrho \qquad (2.23\mathrm{c})$$

$$\mathrm{div}\,\vec{B} = 0, \qquad (2.23\mathrm{d})$$

where only the quasi-static part of the displacement current is incorporated in (2.23b). The induced part of (2.22a) has been neglected. In addition, those charges belonging to the induced part of the electric flux density are not accounted for, from which follows (2.23c). It should be mentioned that (2.22b) is no longer valid for the approximated set of equations in (2.23). Instead, the new system of underlying equations (2.23) is completed by

$$\vec{E} = -\,\mathrm{grad}\,\Phi - \frac{\partial \vec{A}}{\partial t}, \quad \vec{E} = \vec{E}_0 + \vec{E}_\mathrm{i}, \quad \vec{E}_0 = -\,\mathrm{grad}\,\Phi, \quad \vec{E}_\mathrm{i} = -\frac{\partial \vec{A}}{\partial t}, \qquad (2.23\mathrm{e})$$

$$\vec{D} = \vec{D}_0 + \varepsilon \vec{E}_\mathrm{i}, \quad \vec{D}_0 = \varepsilon \vec{E}_0 = \varepsilon_0 \vec{E}_0 + \vec{P}_0, \quad \varrho^{\mathrm{P}} = -\,\mathrm{div}\,\vec{P}_0, \quad \vec{J}^{\mathrm{P}} = \frac{\partial \vec{P}_0}{\partial t}, \qquad (2.23\mathrm{f})$$

$$\vec{B} = \mathrm{curl}\,\vec{A}, \qquad \mathrm{div}\,\vec{A} = -\mu_0 \varepsilon_0 \frac{\partial \Phi}{\partial t}, \quad \vec{H} = \frac{\vec{B}}{\mu_0} - \vec{M}, \quad \vec{J}^{\mathrm{M}} = \mathrm{curl}\,\vec{M}, \qquad (2.23\mathrm{g})$$

$$\vec{J} = \kappa \vec{E}. \qquad (2.23\mathrm{h})$$

Chapter 2 Classical Electrodynamics

When using the new set of equations (2.23), following the procedure of section 2.2 and applying the Lorenz gauge, the wave equations of (2.7) simplify to the equations

$$\Delta \Phi = -\frac{1}{\varepsilon_0} \varrho_{\text{tot}}, \tag{2.24a}$$

$$\Delta \vec{A} = -\mu_0 \vec{J}_{\text{tot}}, \tag{2.24b}$$

where the currents and charges are abbreviated as before with

$$\varrho_{\text{tot}} = \varrho + \varrho^{\text{P}}, \tag{2.24c}$$

$$\vec{J}_{\text{tot}} = \vec{J} + \vec{J}^{\text{P}} + \vec{J}^{\text{M}}. \tag{2.24d}$$

The continuity equation (2.8) remains unchanged. As desired, equations (2.24) fulfill the quasi-stationary condition because the parts causing radiation do no longer exist.

In contrast to the formulations in [36, eq. (42)] and [37, eq. (24)], the additional current part composed of grad Φ is not existent in the new formulation as desired. This has been achieved by applying the different gauge and additionally neglecting the induced charge density in (2.23c). In the formulation based on the Helmholtz decomposition and the Coulomb gauge, this charge density is zero by definition as it belongs to the induced electric flux density which is divergence-free.[6] Consequently, the LQS formulation is more approximative.

As already suggested before, the derived equations (2.24) are not new in terms of practical applicability because the same result is also obtained by letting c_0 approach infinity in (2.7).[7] Often, this approximation is referred to as the Quasi-Static (QS) [41, 42] or the Electro-Magneto-Quasi-Static (EMQS) regime in [43, 44]. The benefit that comes along with the new LQS formulation is a better understanding and a deeper insight into the underlying set of equations (2.23) which can be illustrated by a simple example:

When analyzing the new set of equations, it is possible to additionally neglect the polarization part of the displacement current which changes (2.23b) to

$$\operatorname{curl} \vec{H} = \varepsilon_0 \frac{\partial \vec{E}_0}{\partial t} + \vec{J}. \tag{2.25}$$

This new approximation is useful for low-frequency systems including dielectrics because as a consequence of (2.25), the polarization current density $\vec{J}^{\text{P}}$ is no longer present in (2.24d) and the dielectric influence is only incorporated in the system via ϱ^{P}. One of

---

[6]This is not necessarily the case in inhomogeneous medium where generally not both $\vec{E}_i$ and $\vec{D}_i$ are solenoidal.

[7]This should not be confounded with the static case because the continuity equation (2.8) still couples both equations.

2.4 Quasi-stationary Approximations

the main advantages of this formulation is the fact that the electric part of the system is fully compatible with electrostatic solvers because the electric potential in (2.24a) reacts instantaneously to a change of the charge density. This allows the application of standard methods for electrostatic problems such as the Equivalent Charge Formulation (ECF) for piecewise homogeneous dielectrics [45] or the method of images [46] for a two-layer substrate. In addition, meshing techniques and other findings of the electrostatic MoM technique can be used. The only difference to the static case is the coupling with the magnetic part of the system via the continuity equation (2.8). Sometimes this technique is also referred to as a coupling of MoM and PEEC as in [47] which is shown here to be covered by the LQS formulation.

The magnetic part on the other hand differs from the MQS formulation in the presence of magnetic material. This is due to the current part $\vec{J}^{\mathrm{M}}$ of (2.24d) that is of volumetric nature in general. As already shown in (2.15), the magnetization current is not restricted to the surfaces of homogeneous materials even if the magnetic material has zero conductivity. This is due to the part of the displacement current that has not been neglected in (2.23b) which is different to the MQS case where the total displacement current is being neglected. Despite of this fact, it might be legitimate to neglect the magnetization volume current and to use a surface magnetization current only as the gain of simulation speed outweighs the decreased accuracy of the results.

To the end of this section, a few hints about the range of validity will be presented. The general condition is that the influence of the retardation terms that have been neglected has to be very small [35]. An exact equation is generally very difficult to determine and a rule of thumb is given by [36]

$$\omega_{\max} \ll \frac{c_0}{\max|\vec{r}-\vec{r}'|}, \quad \text{or equivalently} \quad \max|\vec{r}-\vec{r}'| \ll \frac{\lambda_{\min}}{2\pi}, \quad (2.26)$$

where $\max|\vec{r}-\vec{r}'|$ indicates the maximum distance of two points inside the spatial domain and $\lambda_{\min} = c_0/f_{\max}$ is the minimum wavelength.[8] In the case of linear materials involved, the speed of light in (2.26) can be replaced by the lowest medium speed of light of the considered materials. More investigations about the range of validity can be found in [33, 34].

2.4.2 Magneto-Quasi-Static Formulation

In this section, the equations of the MQS formulation are derived. In this approximation, the complete displacement current $\partial \vec{D}/\partial t$ in (2.23b) is neglected which results in Ampere's law

$$\operatorname{curl} \vec{H} = \vec{J}. \quad (2.27)$$

---
[8]Equation (2.26) can be derived by comparing the Green's functions of (2.32).

Chapter 2 Classical Electrodynamics

Repeating the formulation of the magnetic vector potential by substituting (2.3a) and (2.2b) in (2.27) yields

$$\Delta \vec{A} = \mu_0 \vec{J}_{\text{tot}}, \tag{2.28a}$$

where the total current density is composed of two parts in this case according to

$$\vec{J}_{\text{tot}} = \vec{J} + \vec{J}^{\text{M}}. \tag{2.28b}$$

The above equation (2.28) is derived by using the Coulomb gauge $\operatorname{div} \vec{A} = 0$. The MQS continuity equation is directly obtained by (2.27) as

$$\operatorname{div} \vec{J} = 0. \tag{2.29}$$

2.5 Green's Function Method

In this section, solutions to the potential equations will be given by means of the Green's function method. The Green's function can be regarded as representing a solution of the underlying inhomogeneous linear differential equation for a point source whereas the solution to an arbitrary source is given in form of an integral equation. In order to present the solutions to the potentials by means of the Green's function, the wave equations (2.7) and their LQS and MQS counterparts (2.24) and (2.28) are repeated in frequency domain as

Full-wave: *LQS:*   *MQS:*

$$\Delta \underline{\Phi} + k^2 \underline{\Phi} = -\frac{1}{\varepsilon_0} \underline{\varrho}_{\text{tot}}, \qquad \Delta \underline{\Phi} = -\frac{1}{\varepsilon_0} \underline{\varrho}_{\text{tot}}, \tag{2.30a}$$

$$\Delta \underline{\vec{A}} + k^2 \underline{\vec{A}} = -\mu_0 \underline{\vec{J}}_{\text{tot}}, \qquad \Delta \underline{\vec{A}} = -\mu_0 \underline{\vec{J}}_{\text{tot}}, \qquad \Delta \underline{\vec{A}} = -\mu_0 \underline{\vec{J}}_{\text{tot}}, \tag{2.30b}$$

where the total current and charge densities are given in each case as[9]

$$\underline{\varrho}_{\text{tot}} = \underline{\varrho} + \underline{\varrho}^{\text{P}}, \qquad \underline{\varrho}_{\text{tot}} = \underline{\varrho} + \underline{\varrho}^{\text{P}}, \tag{2.30c}$$

$$\underline{\vec{J}}_{\text{tot}} = \underline{\vec{J}} + \underline{\vec{J}}^{\text{P}} + \underline{\vec{J}}^{\text{M}}, \qquad \underline{\vec{J}}_{\text{tot}} = \underline{\vec{J}} + \underline{\vec{J}}^{\text{M}}, \qquad \underline{\vec{J}}_{\text{tot}} = \underline{\vec{J}} + \underline{\vec{J}}^{\text{M}}. \tag{2.30d}$$

In (2.30), the wave number k has been introduced as

$$k = \frac{\omega}{c_0} = \frac{2\pi}{\lambda}. \tag{2.31}$$

---

[9] In contrast to (2.24d), the total current density of the LQS model does not include the polarization current density. This is due to the fact that the polarization displacement current according to (2.25) is additionally neglected.

2.6 Mixed Potential Integral Equation

The left parts of (2.30a) and (2.30b) are differential equations of Helmholtz type whereas the two right parts are of Poisson type, respectively. The solution to the above equations can be found by means of the Green's function which is formulated in frequency domain in case of free space with ε_0 and μ_0 as

Full-wave: $\qquad\qquad\qquad\qquad$ *LQS and MQS:*

$$\underline{G}(\vec{r},\vec{r}') = \frac{1}{4\pi} \frac{e^{-jk|\vec{r}-\vec{r}'|}}{|\vec{r}-\vec{r}'|}, \qquad \hat{G}(\vec{r},\vec{r}') = \frac{1}{4\pi} \frac{1}{|\vec{r}-\vec{r}'|}, \qquad (2.32)$$

with $\vec{r}$ being the observation and $\vec{r}'$ the source point. The left part of (2.32) corresponds to the Helmholtz equation and is a complex valued function because the retardation is transformed to a phase term in the frequency domain. The right part complies with the Poisson equation and does not incorporate the phase term as it reacts instantaneously to the sources for the entire space. In order to unify the following equations, the synonym $G(\vec{r},\vec{r}')$ describes the general Green's function and must be substituted by either $\underline{G}(\vec{r},\vec{r}')$ or $\hat{G}(\vec{r},\vec{r}')$ depending on whether the full-wave solution or the LQS counterpart is used. It should be noted that the above Green's functions of free space can still be applied if dielectric and magnetic materials are accounted for by polarization and magnetization charges and currents.

The Green's function methodology allows for formulating the solution of the potentials of (2.30) as integral equations

$$\Phi(\vec{r}) = \frac{1}{\varepsilon_0} \int_{V'} \varrho_{\text{tot}}(\vec{r}')\, G(\vec{r},\vec{r}')\, \mathrm{d}V', \qquad (2.33\text{a})$$

$$\vec{A}(\vec{r}) = \mu_0 \int_{V'} \vec{J}_{\text{tot}}(\vec{r}')\, G(\vec{r},\vec{r}')\, \mathrm{d}V', \qquad (2.33\text{b})$$

in which $\mathrm{d}V'$ indicates the volume integration over the source vector $\vec{r}'$ located inside the volume V'. By means of (2.33), the potentials can be computed when the charge and current densities are known.

2.6 Mixed Potential Integral Equation

This section will lead to an integral equation based formulation of an EM interconnection system composed of conducting wires such as sketched in Figure 2.1. The basic idea is to express the electric field inside the conductors with Ohm's law (2.16) as a function of the potentials (2.3b) which are in turn substituted by the charges and currents (2.33). Combining this with the continuity equation, the whole system can be expressed for

Chapter 2 Classical Electrodynamics

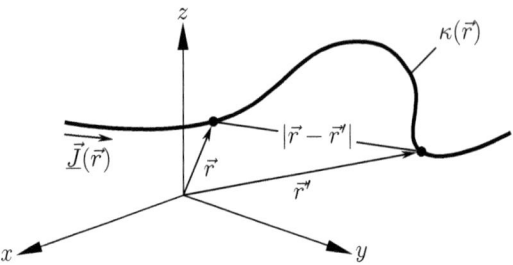

Figure 2.1: Excerpt of an interconnection structure where both the source point $\vec{r}'$ and the observation point $\vec{r}$ are located inside the conductor.

observation points located inside the conductive material according to

$$\frac{\vec{J}(\vec{r})}{\kappa(\vec{r})} + j\omega\mu_0 \int_{V'} \vec{J}_{\text{tot}}(\vec{r}') \, G(\vec{r},\vec{r}') \, \mathrm{d}V' + \operatorname{grad} \underline{\Phi}(\vec{r}) = 0, \qquad (2.34\text{a})$$

$$\frac{1}{\varepsilon_0} \int_{V'} \underline{\varrho}_{\text{tot}}(\vec{r}') \, G(\vec{r},\vec{r}') \, \mathrm{d}V' = \underline{\Phi}(\vec{r}), \qquad (2.34\text{b})$$

$$\operatorname{div} \vec{\underline{J}}_{\text{tot}}(\vec{r}) + j\omega \underline{\varrho}_{\text{tot}}(\vec{r}) = 0. \qquad (2.34\text{c})$$

Since both potentials Φ an $\vec{A}$ are used in the same equation (2.34a), this methodology is called Mixed Potential Integral Equation (MPIE). Instead of MPIE, some contributions use the wording Electric Field Integral Equation (EFIE) as a synonym.

In (2.34), the charges, currents and potentials are the state variables rather than the field quantities. A MoM conform discretization of this system will build the base for the PEEC formulation in chapter 4. This can already be guessed by identifying a resistive term in the first term of (2.34a), an inductive in the second term of (2.34a) as well as a capacitive part in (2.34b). It will be shown in chapter 4 that (2.34a) and (2.34b) can be transformed to the Kirchhoff's Voltage Law (KVL) and (2.34c) to the Kirchhoff's Current Law (KCL), respectively.

If the particular problem does not include any dielectric or magnetic materials, it follows that $\vec{\underline{J}}_{\text{tot}} = \vec{\underline{J}}$ and $\underline{\varrho}_{\text{tot}} = \underline{\varrho}$ and the above system (2.34) describes the electromagnetic behavior completely. Otherwise, the dependencies of the additional current and charge parts on the electric and magnetic fields must be incorporated and solved simultaneously.

If the MQS system is regarded, (2.34c) changes to $\operatorname{div} \vec{\underline{J}}_{\text{tot}} = 0$ and (2.34b) is no longer needed because the charges are not coupled with the currents via (2.34c). The charges do generally not need to be regarded in MQS systems as they do not influence the physical behavior [33].

2.7 Poynting's Theorem

In this section, the network elements resistance, inductance and capacitance are defined in the most general way. These definitions are important for the PEEC method (s. chapter 4) since the circuit elements will be recognized in a modified, discrete form. As a consequence, the network elements will be called partial network elements in the PEEC method. A practicable way to define the resistance, inductance and capacitance is via the different parts of the energy that exist in every non-trivial EM system.

The identification of the different parts of energy can be achieved by using Poynting's theorem which is written for linear and isotropic media in differential form as

$$\frac{\partial}{\partial t}\bigg(\underbrace{\frac{1}{2}\vec{E}\cdot\vec{D}}_{w_\mathrm{e}}+\underbrace{\frac{1}{2}\vec{B}\cdot\vec{H}}_{w_\mathrm{m}}\bigg) = -\mathrm{div}\big(\underbrace{\vec{E}\times\vec{H}}_{\vec{S}}\big) - \vec{E}\cdot\vec{J}. \qquad (2.35)$$

This equation is obtained by building the divergence of $\vec{E}\times\vec{H}$, using the vector relation $\mathrm{div}\,(\vec{E}\times\vec{H}) = \vec{H}\cdot\mathrm{curl}\,\vec{E} - \vec{E}\cdot\mathrm{curl}\,\vec{H}$, substituting (2.1a) and (2.1b) and making use of the identities $2\vec{E}\cdot\partial/\partial t\vec{D} = \partial/\partial t(\vec{E}\cdot\vec{D})$ and $2\vec{H}\cdot\partial/\partial t\vec{B} = \partial/\partial t(\vec{B}\cdot\vec{H})$ that are valid for linear and isotropic materials. In (2.35), the abbreviated quantities are the electric energy density w_e, the magnetic energy density w_m and the Poynting vector $\vec{S}(\vec{r},t)$. The equation can be interpreted as an energy conservation law:

The change of the energy density being stored in the electric and magnetic fields can be accounted for by the two parts in the right hand side of (2.35). The first one describes the electromagnetic energy flow characterized by the Poynting vector whereas the second term represents dissipation in form of Joule heating. By integrating (2.35) over a specific volume V, the electric energy W_e, magnetic energy W_m and ohmic losses P_l are obtained as

$$W_\mathrm{e} = \frac{1}{2}\int_V \vec{E}\cdot\vec{D}\,\mathrm{d}V, \qquad W_\mathrm{m} = \frac{1}{2}\int_V \vec{H}\cdot\vec{B}\,\mathrm{d}V, \qquad P_\mathrm{l} = \int_V \vec{E}\cdot\vec{J}\,\mathrm{d}V. \qquad (2.36)$$

The three parts of the above equation can be used to derive definitions for the resistance, inductance and capacitance, which will be focused on in the following three subsections.

2.7.1 Definition of Resistance

In order to derive an expression for the resistance, a conductor with the volume V but not necessarily uniform cross section A and the conductivity $\kappa > 0$ needs to be defined according to Figure 2.2. The integral of the current density over an arbitrary cross section yields the total current $I = \int \vec{J}\cdot\mathrm{d}\vec{A}$ which is regarded as the excitation of the system. In this case, the resistance R of the conductor is given by the losses P_l of (2.36) normalized to the square of the total current which can be expressed as

$$P_1 = RI^2. \tag{2.37}$$

Substituting P_1 of (2.36) and the electric field by Ohm's law (2.16) inside the conductor yields

$$R = \frac{1}{I^2} \int_V \frac{|\vec{J}(\vec{r})|^2}{\kappa(\vec{r})} \, dV. \tag{2.38}$$

If the cross section and the conductivity do not change with the length l, the per-unit-length resistance R' can be introduced by replacing the volume element by $dV = dA\,dl$ as

$$R' = \frac{R}{l} = \frac{1}{I^2} \int_A \frac{|\vec{J}(\vec{r})|^2}{\kappa(\vec{r})} \, dA. \tag{2.39}$$

In the case of a homogeneous conductivity κ and current density $|\vec{J}| = I/A$ which is the case for stationary problems, the DC resistance R_{DC} of a conductor with cross section A and length l is given by the well known expression

$$R_{\text{DC}} = \frac{l}{\kappa A}. \tag{2.40}$$

2.7.2 Definition of Inductance

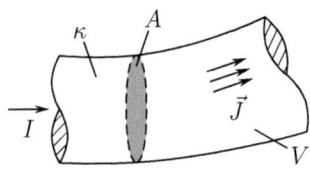

Figure 2.2: Conductor with an arbitrary volume

The most simple case to define an inductance L is again the single conductor setup displayed in Figure 2.2 which is located in free space. The conductor is excited with the current I. In this case, the inductance can be defined as the stored magnetic energy normalized to the square of the current with the relation

$$W_{\text{m}} = \frac{1}{2} L I^2. \tag{2.41}$$

Combining this equation with the magnetic energy definition (2.36), the most general formulation for the inductance is obtained as

$$L = \frac{1}{I^2} \int_V \vec{H} \cdot \vec{B} \, dV. \tag{2.42}$$

Since $\vec{H}$ and $\vec{B}$ are generally defined in the whole space, the above equation allows for separating the integration volume into several regions. If the volume is restricted to the

2.7 Poynting's Theorem

interior of the conductor, the solution of (2.42) is called internal inductance L_{int}. The other way around, the integral of the magnetic energy outside the conductor is referred to as external inductance L_{ext}. If the integration volume is extended over the whole space, the related inductance is called total inductance.

There exists a different expression for the total inductance which is obtained by substituting the magnetic flux density of the integral in (2.42) by $\vec{B} = \operatorname{curl} \vec{A}$ of (2.3a) and applying the vector identity $\vec{H} \cdot \operatorname{curl} \vec{A} = \operatorname{div}(\vec{A} \times \vec{H}) + \vec{A} \cdot \operatorname{curl} \vec{H}$ for the resulting expression. The volume integral over the first term $\operatorname{div}(\vec{A} \times \vec{H})$ can be converted via Gauss' theorem to a surface integral which in turn can be shown to vanish for an infinite surface, e.g. [32, p. 341]. In the second term, the substitution of $\operatorname{curl} \vec{H} = \vec{J}$ can be applied for the stationary or MQS cases, from which the total magnetic energy can be expressed as

$$W_{\text{m}} = \frac{1}{2} \int_V \vec{A} \cdot \vec{J} \, \mathrm{d}V. \tag{2.43}$$

Substituting the magnetic vector potential by the solution of free space (2.33b), the total inductance is obtained by inserting the above equation in (2.41) which results in

$$L = \frac{\mu_0}{4\pi I^2} \int_V \int_{V'} \frac{\vec{J}(\vec{r}) \cdot \vec{J}(\vec{r}')}{|\vec{r} - \vec{r}'|} \, \mathrm{d}V' \, \mathrm{d}V. \tag{2.44}$$

Note that the volume integrations are now performed over the conducting region only where the current density differs from zero. The separation of internal and external inductance is no longer possible by this equation as it is based on the magnetic energy stored in the whole space.

The extension to multi-conductor systems is straightforward and allows the introduction of the mutual inductance L_{mn} between two conductors m and n as

$$L_{mn} = \frac{\mu_0}{4\pi I_m I_n} \int_V \int_{V'} \frac{\vec{J}_m(\vec{r}) \cdot \vec{J}_n(\vec{r}')}{|\vec{r} - \vec{r}'|} \, \mathrm{d}V' \, \mathrm{d}V. \tag{2.45}$$

As can be seen from the above equation, $L_{mn} = L_{nm}$ from which follows that an introduced inductance matrix is symmetric. In addition, from (2.45) it can be concluded that two currents which are flowing perpendicular to each other do not share a mutual inductance.

For reasons of completeness it should be mentioned that it is also possible to define a per-unit-length inductance as $L' = L/l$ for some 2D applications.

Inductances of closed loops A further inductance definition which is based on the magnetic flux Ψ can be given for closed current loops though this definition has more

Chapter 2 Classical Electrodynamics

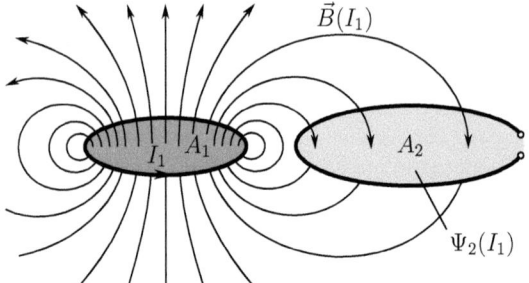

Figure 2.3: Visualization of the inductance concept via magnetic fluxes. The mutual inductance is defined as the magnetic flux in the secondary coil Ψ_2 divided by the primary current I_1.

restrictions than the equations above. Validity is given only for closed conductor loops with neglected internal inductance L_int under MQS assumptions. For the derivation of the inductance by means of the magnetic flux, in (2.42) a magnetic scalar potential for the non-current-carrying regions is introduced. After some mathematical intermediate steps as explained in [32, p. 345] for instance, the inductance can be expressed as

$$L = \frac{1}{I}\Psi = \frac{1}{I}\int_A \vec{B} \cdot \mathrm{d}\vec{A}, \tag{2.46}$$

in which A is the area bounded by the current loop. The above equation states that the inductance is given by the magnetic flux through the coil area normalized to the current through the conductor. It should be mentioned again that the internal inductance L_int is not included in (2.46).

If the current loop is composed of N_turn windings, the inductance is increased by the factor of N_turn^2. This matter of fact can be verified by evaluating (2.44) with an N_turn-times higher current density but normalizing to the current of a single winding only. This rule is exact only if all windings are positioned at the same location. Especially for spiral coils this is no more than a rule of thumb since the size of the turns decreases towards the interior of the coil.

The extension to multi-inductor systems is visualized in Figure 2.3. By indexing the different current loops via the subscript i, (2.46) can be written for the mutual inductance L_{mn} between loop m and n as

$$L_{mn} = \frac{1}{I_n} \int_{S_q} \vec{B} \cdot \mathrm{d}\vec{A}_m \bigg|_{I_i = 0,\ i \neq n}. \tag{2.47}$$

If $m = n$, the result of (2.47) is named self-inductance and is often written as L_m instead of L_{mm}.

For some applications it is reasonable to normalize the mutual inductance by the geometric mean value of the self-inductances, which results in the so-called coupling factor or coupling coefficient

$$k_{mn} = \frac{L_{mn}}{\sqrt{L_m L_n}}. \tag{2.48}$$

In case of a two coil arrangement, the coupling factor k is usually written without subscripts and the mutual inductance is abbreviated with M. As can be seen from (2.48), the absolute value of the coupling factor ranges in between 0 and 1 whereas $k = 0$ is the case for zero coupling and $k = 1$ for two identical coils that are located at the same spatial position and orientation.

2.7.3 Definition of Capacitance

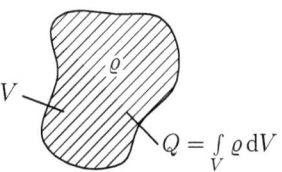

Figure 2.4: Charge density distributed in free space

According to the derivation of the resistance and inductance, the basis for defining the capacitance is a volume with an arbitrary charge density $\varrho(\vec{r})$ distributed in free space. The total charge Q is given by the volume integral over the charge density as presented in Figure 2.4. According to the magnetic case, the electric energy of (2.36) can be normalized to the square of the total charge as

$$W_{\mathrm{e}} = \frac{1}{2} P Q^2, \tag{2.49}$$

in which P is the coefficient of potential defined by

$$P = \frac{1}{Q^2} \int_V \vec{E} \cdot \vec{D} \, \mathrm{d}V. \tag{2.50}$$

It will be seen that the coefficient of potential is the reciprocal value of the capacitance if a conductor exists. Note that this is not necessarily the case here. According to the magnetic energy from the last section, the total electric energy can be expressed in a different form under the conditions that static or EQS formulations are valid, the reference potential $\Phi = 0$ is at infinity and the integration is performed over the whole space. In this case, the total electric energy can be converted to

$$W_{\mathrm{e}} = \frac{1}{2} \int_V \varrho \Phi \, \mathrm{d}V, \tag{2.51}$$

which is achieved by inserting $\vec{E} = -\operatorname{grad} \Phi$ to the first part of (2.36) and using the vector identity $\vec{D} \cdot \operatorname{grad} \Phi = \operatorname{div}(\Phi \vec{D}) - \Phi \operatorname{div} \vec{D}$. The volume integral over the first term $\operatorname{div}(\Phi \vec{D})$ can be converted via Gauss' theorem to the surface integral which in turn can be shown to vanish for an infinite surface, e.g. [32, p. 114]. In the second term, $\operatorname{div} \vec{D} = \varrho$ can be substituted which allows for writing the total electric energy according to (2.51). When replacing the potential in (2.51) by the free space solution of (2.33a) and substituting the electric energy to (2.49), a formulation for the coefficient of potential similar to (2.44) is achieved

$$P = \frac{1}{4\pi\varepsilon_0 Q^2} \int_V \int_{V'} \frac{\varrho(\vec{r})\varrho(\vec{r}')}{|\vec{r}-\vec{r}'|} \,\mathrm{d}V' \,\mathrm{d}V. \tag{2.52}$$

In order to define a capacitance, the setup shown in Figure 2.4 is modified in such a way that the charged region is superposed by a conductor which is raised to the constant potential Φ_0 with the same total charge Q as before (s. Figure 2.5). The voltage w.r.t. the reference potential is $U = \Phi_0$ because the reference potential at infinity is zero by definition. For this conductor, the voltage rather than the charge can be regarded as the source and normalizing the energy by the square of the voltage yields

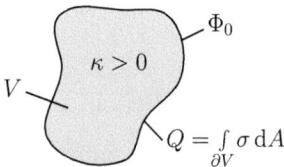

Figure 2.5: Charged conductor in free space

$$W_\mathrm{e} = \frac{1}{2}CU^2, \tag{2.53}$$

in which C is the capacitance defined by

$$C = \frac{1}{U^2} \int_V \vec{E} \cdot \vec{D} \,\mathrm{d}V. \tag{2.54}$$

Note that the definitions (2.49) to (2.52) still hold due to the charge density on the conducting material. Because the potential is assumed to be constant inside and on the surface of the conducting material, the expression of (2.51) can be simplified to $W_\mathrm{e} = 1/2\,U\,Q$. This is because the potential $\Phi = U$ can be taken outside of the integral and the remaining integral yields the total charge. Equaling the obtained expression for the electric energy with the definitions in (2.49) and (2.53) results in the standard relations

$$C = \frac{Q}{U}, \quad \text{and} \quad C = \frac{1}{P}, \tag{2.55}$$

which confirm the reciprocal relationship of the capacitance and the coefficient of potential. The above equations in combination with (2.33a) allow for an alternative solution to (2.52) for conductors with a constant potential

$$P = \frac{\Phi(\vec{r}_i)}{Q} = \frac{1}{4\pi\varepsilon_0 Q} \int_{V'} \frac{\varrho(\vec{r}')}{|\vec{r}_i - \vec{r}'|} \,\mathrm{d}V', \tag{2.56}$$

in which $\vec{r}_i$ is an arbitrary vector inside or on the surface of the conductor.

For multiconductor systems, two different types of capacitance matrices can be derived, depending on whether the absolute potentials or voltages are referred to.

2.7 Poynting's Theorem

2.7.4 Definition of Impedance

In this subsection, the complex impedance of a one-port black box network is introduced by evaluating the Poynting's theorem in frequency domain.[10] Repeating the steps of (2.35) with $\vec{E} \times \vec{H}^*$, integrating over the volume V and reordering terms yields the complex Poynting's theorem

$$\underline{P} = -\int_{\partial V} \left(\vec{\underline{E}} \times \vec{\underline{H}}^*\right) \cdot \mathrm{d}\vec{A} = \underbrace{\int_V \vec{\underline{E}} \cdot \vec{\underline{J}}^* \, \mathrm{d}V}_{P_\mathrm{l}} + 2j\omega \left[\underbrace{\frac{1}{2}\int_V \vec{\underline{B}} \cdot \vec{\underline{H}}^* \, \mathrm{d}V}_{W_\mathrm{m}} - \underbrace{\frac{1}{2}\int_V \vec{\underline{E}} \cdot \vec{\underline{D}}^* \, \mathrm{d}V}_{W_\mathrm{e}}\right],$$
(2.57)

in which all complex field quantities are written as Root Mean Square (RMS) values in order to avoid the scaling factor $1/2$ and to be consistent with the energy definitions of (2.36). In (2.57), the introduced complex power $\underline{P}$ describes the power delivered to the one-port network and is identical to the negative power flow characterized by the complex Poynting vector.

The relation (2.57) allows the definition of an impedance $\underline{Z}$ by normalizing the complex power to the square of the absolute RMS value of the current which is flowing into the network as [48]

$$\underline{Z} = \frac{\underline{P}}{|\underline{I}|^2} = \frac{P_\mathrm{l} + 2j\omega(W_\mathrm{m} - W_\mathrm{e})}{|\underline{I}|^2} = R + jX.$$
(2.58)

In the last equality, the impedance has been decomposed into real and imaginary parts. The real part is identical to the resistance R because of (2.37) and the imaginary part is named reactance X. If the electric energy of the one-port network can be neglected, e.g. when using the MQS approach, a simplified version of (2.58) can be derived by substituting the magnetic energy by (2.41) resulting in

$$\underline{Z} = R + j\omega L.$$
(2.59)

In this case, the impedance can be written in terms of a series connection of a resistance and an inductance.

---

[10] In this model, radiation is not regarded. A more general definition is presented in [31].

Chapter 3
Inductive Power Transmission

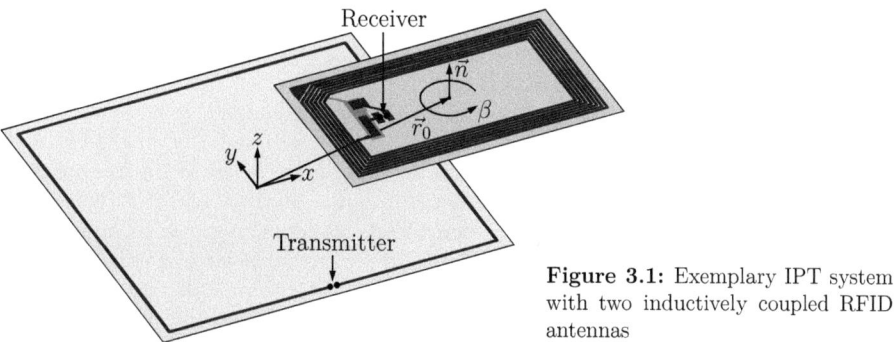

Figure 3.1: Exemplary IPT system with two inductively coupled RFID antennas

In this chapter, the fundamentals of Inductive Power Transfer (IPT) systems are discussed. In general, an IPT system is used to transmit power from a source to a remote receiver wirelessly as visualized in Figure 3.1 for the case of an RFID system. The functionality is as follows: The source drives a current-carrying conductor which generates a magnetic field according to Ampere's law (2.1b). If the magnetic field changes over time, an electric field is induced by Faraday's law (2.1a) which causes a voltage drop in an attendant second conductor. If this conductor is located in a receiver positioned remotely to the source, the induced voltage can be rectified and afterwards be used for powering the device. In most practical applications, both conductors are designed as closed loops whereas the specific shape, turn configuration etc. may differ for each application. The current loops are commonly named coils, inductors or antennas.

This chapter first derives the fundamental relations of IPT systems by analyzing a circular loop antenna. For this special kind of antenna, exact analytical expressions can be given, e.g. for the electromagnetic fields and the inductance. Approximating the loop antenna with the magnetic dipole formulation allows for introducing near- and far-fields as well as different loss mechanisms. The second section of this chapter is focused on general design aspects of IPT systems such as the appropriate frequency range, different quality factor definitions as well as the geometrical layout of a rectangular Printed Spiral Coil (PSC). The third section concentrates on a network description of IPT systems whereas different macromodels of the individual antennas are coupled via the transformer concept. The entire IPT system design is analyzed in terms of efficiency maximization and field emission minimization.

3.1 Small Circular Loop Antenna

In order to derive the physical relevant properties of IPT systems such as near- and far-fields, directivity, wave propagation, inductance as well as resistive and radiative losses, a single loop antenna is regarded in this section. This is because a closed current loop is the fundamental device of almost every inductive system. The most simple loop antenna setup which can be analyzed by exact analytical equations is the loop antenna of circular shape, small electrical size (constant current assumed) and negligible cross section of the wire. The following analysis is in addition restricted to the free space or a homogeneous medium in general. If a detailed and accurate analysis of the specific behavior of more complex structures is demanded, numerical approaches such as the PEEC method can be applied. Nevertheless, the fundamental laws of physics do not change.

In the following, the circular loop antenna displayed in Figure 3.2 with the driving current $\underline{I}_0$ is considered. Although this setup might seem quite simple, the exact formulations for the electric and magnetic fields in the whole space are mathematically not trivial to handle. One possibility is to described the fields in terms of a double series representation as will be detailed in the following. The loop antenna has been of interest for researchers for many years. An analysis including the region in the close proximity of the loop has been presented by Werner in 1996 [49]. The obtained results for the special case in which the current is uniformly distributed over the circumference of the loop will be reviewed in the following paragraph.

Vector Potential and Field Components The circular loop antenna according to Figure 3.2 with the radius r_0 is analyzed at an arbitrary field point $\vec{r}$ in spherical coordinates characterized by r, ϑ and φ. The electromagnetic fields are expressed via the magnetic vector potential which can be written as an integral in (2.33b) for the case that the current is given. In the case of an infinitely thin conductor, the volume integral of (2.33b) reduces to a line integral over the contour of the loop. If the diameter of the loop is small compared to the wavelength, the current can be regarded as constant $\underline{I} = \underline{I}_0$ and the solution does not depend on the azimuthal angle φ. In this case, the vector potential of (2.33b) is composed of a φ-component only. Due to the azimuthal symmetry, it is sufficient to analyze the vector potential at $\varphi = 0$ for example, resulting

3.1 Small Circular Loop Antenna

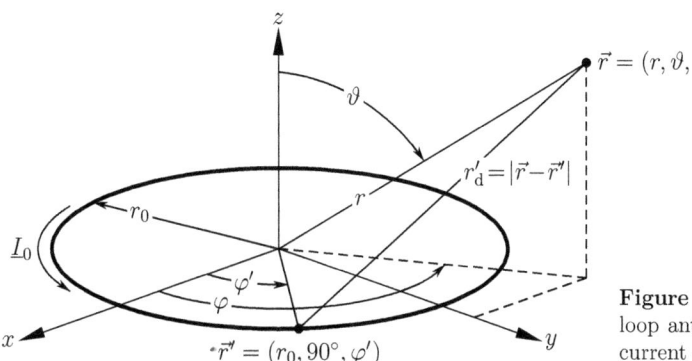

Figure 3.2: Thin circular loop antenna with a uniform current excitation

in the following expression

$$\underline{A}_\varphi(r,\vartheta) = \frac{\mu_0 r_0 \underline{I}_0}{2\pi} \int_0^\pi \cos\varphi' \frac{e^{-jkr'_\mathrm{d}}}{r'_\mathrm{d}} \mathrm{d}\varphi', \tag{3.1}$$

where the following abbreviations have been introduced

$$r_\mathrm{d} = \sqrt{r^2 + r_0^2}, \qquad r'_\mathrm{d} = |\vec{r} - \vec{r}\,'| = \sqrt{r_\mathrm{d}^2 - 2r_0 r \sin\vartheta \cos\varphi'}. \tag{3.2}$$

The integral of (3.1) can be solved by expanding the exponential function into a power series and integrating the terms element by element. The solution can be written as [49]

$$\underline{A}_\varphi(r,\vartheta) = \frac{k\mu_0 r_0 \underline{I}_0}{2j} e^{-jkr_\mathrm{d}} \sum_{m=1}^\infty \sum_{n=0}^{2m-1} \underline{D}_{mn} \frac{[(k^2 r_0 r \sin\vartheta)/2]^{2m-1}}{(k\, r_\mathrm{d})^{2m+n}}, \tag{3.3a}$$

with the coefficients

$$\underline{D}_{mn} = \frac{1}{(2j)^n} \frac{(2m+n-1)!}{(2m-n-1)!n!} \frac{(-1)^m}{m!(m-1)!}. \tag{3.3b}$$

The magnetic and electric field components can be computed by building the curl of (3.3a) according to (2.3a) and (2.1a) as $\vec{\underline{H}} = 1/\mu_0\,\mathrm{curl}\,\vec{\underline{A}}$ and $\vec{\underline{E}} = 1/(j\omega\varepsilon_0)\,\mathrm{curl}\,\vec{\underline{H}}$. Due to the fact that the r- and ϑ-components of the vector potential do not exist and the remaining φ-component depends on r and ϑ only, the remaining nonzero components

Chapter 3 Inductive Power Transmission

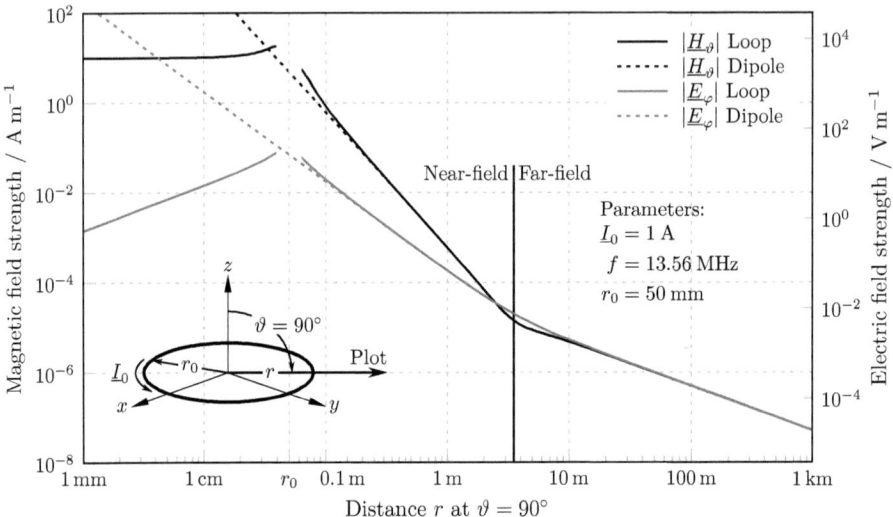

Figure 3.3: Electric and magnetic field components of the loop antenna and the dipole approximation as a function of the radial distance at $\vartheta = 90°$. Since only the first 30 terms of the series (3.4) have been computed, the singularity at r_0 is not captured correctly. For this reason, the field components are not visualized near $r = r_0$. The different y-axes are scaled by the wave impedance of free space in order to obtain an equality of both electric and magnetic field strengths in the far-field domain. In the near-field region, the magnetic field dominates the electric field up to several orders of magnitude w.r.t. the free-space relation.

of the electric and magnetic fields are given by [49]

$$\underline{E}_\varphi(r,\vartheta) = \sqrt{\frac{\mu_0}{\varepsilon_0}} \frac{k^2 r_0 \underline{I}_0}{-2} e^{-jkr_\mathrm{d}} \sum_{m=1}^{\infty}\sum_{n=0}^{2m-1} \underline{D}_{mn} \frac{[(k^2 r_0 r \sin\vartheta)/2]^{2m-1}}{(k\,r_\mathrm{d})^{2m+n}}, \tag{3.4a}$$

$$\underline{H}_r(r,\vartheta) = \frac{k^3 r_0^2 \underline{I}_0 \cos\vartheta}{2j} e^{-jkr_\mathrm{d}} \sum_{m=1}^{\infty}\sum_{n=0}^{2m-1} m\underline{D}_{mn} \frac{[(k^2 r_0 r \sin\vartheta)/2]^{2m-2}}{(k\,r_\mathrm{d})^{2m+n}}, \tag{3.4b}$$

$$\underline{H}_\vartheta(r,\vartheta) = \frac{k^3 r_0^2 \underline{I}_0 \sin\vartheta}{-2j} e^{-jkr_\mathrm{d}} \sum_{m=1}^{\infty}\sum_{n=0}^{2m-1} m\underline{D}_{mn} \frac{[(k^2 r_0 r \sin\vartheta)/2]^{2m-2}}{(k\,r_\mathrm{d})^{2m+n}}$$
$$\cdot \left[1 - \frac{r^2}{r_\mathrm{d}^2}\frac{(2m+n) + jkr_\mathrm{d}}{2m}\right]. \tag{3.4c}$$

In Figure 3.3, the field components are visualized for the radial distance at $\vartheta = 90°$. As can be seen from (3.4b), the radial component $\underline{H}_r$ vanishes at this angle because of $\cos 90° = 0$. The absolute value of the remaining $\underline{H}_\vartheta$ component is almost constant inside the loop whereas the component itself changes the sign at $r = r_0$. The singularity

3.1 Small Circular Loop Antenna

at this point is not captured correctly since only 30 terms of the series in (3.4c) have been computed. The azimuthal electric field component $\underline{E}_\varphi$ drops to zero for $r = 0$ which is due to the symmetry of the loop. For larger r, the electric field increases until $r \to r_0$. For $r > r_0$ it decreases again.

Dipole Approximation For high r/r_0 ratios, an approximation of (3.4) can be derived which greatly reduces the complexity of the equations because it is sufficient to regard the first term $m = 1$ of the outer sum only. When substituting $r_\mathrm{d} \approx r$ in (3.2), the fields of (3.4) reduce to

$$\underline{E}_\varphi(r, \vartheta) \approx \frac{k^3 r_0^2 \underline{I}_0 \sin \vartheta}{4} \left[\frac{1}{kr} - j\frac{1}{(kr)^2} \right] \sqrt{\frac{\mu_0}{\varepsilon_0}} e^{-jkr}, \tag{3.5a}$$

$$\underline{H}_r(r, \vartheta) \approx \frac{k^3 r_0^2 \underline{I}_0 \cos \vartheta}{2} \left[j\frac{1}{(kr)^2} + \frac{1}{(kr)^3} \right] e^{-jkr}, \tag{3.5b}$$

$$\underline{H}_\vartheta(r, \vartheta) \approx \frac{k^3 r_0^2 \underline{I}_0 \sin \vartheta}{4} \left[-\frac{1}{kr} + j\frac{1}{(kr)^2} + \frac{1}{(kr)^3} \right] e^{-jkr}. \tag{3.5c}$$

These equations are equivalent to the field distribution of an elementary magnetic dipole [32]. The results of (3.5) have been included in Figure 3.3 in order to demonstrate the validity of the above approximation for large radial distances. On the other hand, the error for distances in the region of the radius and below increases as expected. The results of the approximation (3.5) are used in the following subsections to deduce important properties of IPT systems.

3.1.1 Near- and Far-field Regions

The slopes of the curves in Figure 3.3 can directly be assigned to the different parts of (3.5). For large distances, the $1/(kr)$ terms dominate. This region is called far-field and only the $\underline{E}_\varphi$ and $\underline{H}_\vartheta$ components exist in there. As will be seen in the next subsection 3.1.2, radiation is dominant only in this spatial domain. One property of the far-field region is the fact that the electric and the magnetic fields have a fixed ratio which can be expressed in terms of the wave impedance of free space Z_0 as

$$Z_0 = \left| \frac{\underline{E}_\varphi}{\underline{H}_\vartheta} \right| = \sqrt{\frac{\mu_0}{\varepsilon_0}} \approx 377\,\Omega. \tag{3.6}$$

Followed by this, the electric and magnetic energy densities of (2.35) are of equal parts in this domain. In order to compare the electric and magnetic fields in Figure 3.3, the right y-axis is scaled by Z_0 which leads to an adjustment of the curves in the far-field region. When decreasing the distance and approaching the loop antenna region, it can be observed that the remaining parts of (3.5) start to outweigh the $1/(kr)$ terms.

Chapter 3 Inductive Power Transmission

The transition border can be obtained by equaling the absolute values of each term $1/(kr) = 1/(kr)^2 = 1/(kr)^3$ of (3.5) which results in $kr = 1$, obviously. This identity leads to the expression for the near-field far-field border

$$r_{\text{NearFar}} = \frac{1}{k} = \frac{\lambda}{2\pi} = \frac{c_0}{\omega}. \tag{3.7}$$

The different regions separated by this characteristic value have been indicated in Figure 3.3. When comparing the result of (3.7) with the range of validity of the quasi-stationary assumption (2.26), an equivalence can be observed when replacing the maximum spatial distance $\max|\vec{r} - \vec{r}'|$ with the border r_{NearFar}. From this follows the important property that the quasi-stationary assumption holds in the near-field region of the loop antenna.

When comparing the different parts of the dipole approximation (3.5) in the near-field, the magnetic field increases with $1/r^3$ when approaching the antenna whereas the electric field of (3.5a) increases with $1/r^2$ only. This can be interpreted as a predominance of the magnetic field in the near-field region w.r.t. the fixed free-space relation of (3.6). In here, the magnetic field dominates the electric counterpart of up to several orders of magnitude depending on the specific parameter settings. For this reason, loop antennas which are operating in the near-field domain are often called inductors or coils and can be analyzed under the MQS assumption.

3.1.2 Resistive and Radiative Losses

In this section, the different loss mechanisms of the loop antenna visualized in Figure 3.2 are discussed. Since IPT systems aim to be operated at high efficiencies, all parts of losses should be minimized. This in turn demands an accurate modeling of the involved loss mechanisms. As can be seen from Poynting's theorem (2.35), in general there exist two different parts that cause unwanted losses.

Radiation Losses The first part is constituted by the radiation losses which can be quantified by building the real part of the complex Poynting vector $\vec{S} = 1/2\,\vec{E} \times \vec{H}^*$. By inserting the expressions of (3.5) and performing algebraic conversions, the only remaining component is the radial one in the far-field domain. In addition, no power is radiated in the z-direction due to the zero electric field in this direction. The radiation losses can be described by a resistance R_S which can be introduced according to (2.38) by integrating the real part of the complex Poynting vector over the surface of a sphere and normalizing to the square of the current resulting in

$$R_S = \left(\frac{2\pi r_0}{\lambda}\right)^4 \frac{\pi}{6} Z_0. \tag{3.8}$$

3.1 Small Circular Loop Antenna

The bracketed term with the power of four indicates that a sufficient radiation exists only if the circumference of the loop $2\pi r_0$ is comparable to the wavelength λ. Contrary, for sufficiently small loop antennas, the radiation resistance R_S is negligible w.r.t. the ohmic losses. Since the radiation losses are unwanted in IPT systems, (3.8) defines an upper limit for the possible frequency range.

As a consequence of integrating the real part of the Poynting vector over the sphere, no power is transferred or radiated in the near-field region[1] except if a consumer is present which changes the overall field distribution. In this case, the real part of the Poynting vector is directed from the source to the receiver.

Resistive Losses In order to determine resistive losses of the loop antenna displayed in Figure 3.2, the conductor must be equipped with a nonzero cross section. The most simple case is again a circular cross section with the radius a as depicted in Figure 3.4. For loop antennas with high r_0/a ratios, the conductor can locally be regarded as a straight line and a homogeneous current density can be concluded for the stationary case. This allows for an approximated expression for the ohmic losses by evaluating the DC resistance from (2.40) with regarding $l = 2\pi r_0$ as

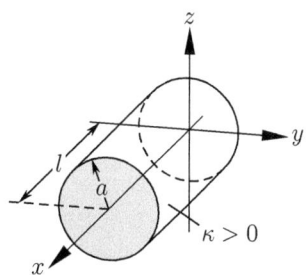

Figure 3.4: Circular conductor

$$R_\mathrm{DC,loop} = \frac{2r_0}{\kappa a^2}. \tag{3.9}$$

The behavior is different at high frequencies in which eddy currents are induced due to the mutual coupling of (2.1a) and (2.1b) as Ohm's law $\vec{E} = \kappa \vec{J}$ holds inside the conductors. When considering the signs of both equations, it can be observed that the current density is expelled from the inside of the conductor towards the boundary. This effect is often called *skin effect* which states that the current tends to flow on the surfaces of the conductors at high frequencies. A common way to describe the effect is to introduce the so-called skin depth δ as

$$\delta = \sqrt{\frac{2}{\omega \mu \kappa}}. \tag{3.10}$$

The skin depth characterizes the depth inside the conducting material in which the field amplitudes have decayed to approximately 37 % of the maximum value at the surface. The expression for the skin depth can be derived by analyzing a conducting half-space under MQS assumptions as presented in [50] for instance. It can further be shown that

---
[1] Assumed that the far-field terms can be neglected.

Chapter 3 Inductive Power Transmission

for a conducting sheet, the total Alternating Current (AC) resistance equals the DC resistance of a virtual setup, consisting of two layers at the boundaries of the sheet with the skin depth of (3.10) as the thickness [50]. Moreover, the assumption of a homogeneous current density inside the skin layer and zero current density elsewhere can be used as an approximation for different shaped conductors. Validity is given when the skin depth is small compared to the cross sectional dimensions and the curvature of the surface can locally be neglected. Exact analytical solutions for the AC current distribution of conductors exist only for simple cross sectional shapes such as circles (s. section 6.1.2) and ellipses. Conductors with rectangular cross section are much more complicated to analyze. In section 6.1.3 on page 116, attempts to find the exact current distribution of conductors with rectangular cross section are presented.

Besides the described skin effect, a further change of the total current distribution inside the conductors due to the eddy currents comes along for arrangements with multiple conductors located in close proximity. This is especially the case in IPT systems in which spiral multi-turn coils are used. The current of each conductor influences the current distributions of all nearby conductors. This behavior is called *proximity effect* and is mathematically more complicated to describe than the skin effect because it is not only a function of the cross section but also of the relative alignment of the conductors. To the author's knowledge, no analytical expressions exist for capturing the proximity effect of a multi-conductor system. As an alternative, numerical modeling techniques with a fine volume mesh of the interior of the conductors can be applied which, however, may cause long simulation time.

3.1.3 Inductance

Due to the dominating magnetic field in the near-field domain of the loop antenna, the stored electric energy can be neglected and the total energy can be expressed in form of the inductance via (2.41). This allows for modeling the complex port impedance of the loop antenna in form of a series connection of a resistance and an inductance according to (2.59).

In order to find expressions for the inductance of the loop antenna according to Figure 3.2, the cross section of the wire has to be nonzero because the magnetic energy and inductance, respectively, would be infinite otherwise. The most simple case is again the circular cross section visualized in Figure 3.4 with a being the radius of the wire. In the stationary case, an expression for the inductance of the loop antenna as presented in Figure 3.2 is given in [51]. In this reference, an exact solution to the total inductance of (2.44) has been found in torus coordinates whereas the equation is again given by means of a cumbersome series expression. In addition to that, an approximated version is derived in [51] as

$$L = \mu_0 r_0 \left[\ln\left(8\frac{r_0}{a}\right) - \frac{7}{4} \right], \qquad (3.11)$$

which can be used as a rough estimation during the design process. As mentioned before, this equation is valid for the stationary case only with a current density being proportional to $1/r$. For increasing frequencies, the skin effect pushes the current towards the boundary which in turn causes the internal as well as the total inductances to decrease. An accurate determination of the frequency-dependent inductance behavior can be performed via numerical computations with a high resolution of the interior of the conductor according to the resistive case.

3.2 Design Constraints

This section concentrates on different design aspects which are needed for practical IPT applications. After presenting important hints for the choice of the working frequency, different definitions of quality factors are introduced. Afterwards, the geometry of a rectangular PSC is focused on as it is a popular device due to its easy producibility. Other mounting forms of inductors such as coils made of wound wires are not treated in this work although the general behavior is equivalent.

3.2.1 Frequency Range

When choosing the appropriate working frequency for a specific IPT application, the following hints should be considered.

A mandatory prerequisite is that legislative limitations are met because of the emission of electromagnetic fields in the proximity of human beings and other electronic devices which could interact with the IPT system. Besides these limitations in terms of maximum field strengths, not all frequencies are allowed for free use. A set of frequency bands that can be used without a special license are the so-called Industrial, Scientific and Medical (ISM) bands. A further discussion of ISM bands is out of the focus of this work. Instead, some efficiency considerations that influence the choice of the operating frequency are presented.

Generally the efficiency of the wireless inductive link increases with rising frequencies, since the inducted voltage increases linearly with the frequency. This can directly be concluded from Faraday's law (2.1a) in the frequency domain. On the other hand, this effect is partially compensated by the fact that the maximum allowed field strengths are higher at lower frequencies [9].

In addition, an upper limit for the suitable frequency range exists for different reasons: First, the near-field far-field border of (3.7) behaves reciprocal to the frequency from which follows that an increased frequency also means a reduced maximum spatial range of functionality. Second, the radiation resistance of (3.8) increases with the frequency to the power of four, which makes an efficient inductive link impossible for frequencies above a specific border. A third natural upper frequency limitation is given by the Self-Resonant Frequency (SRF) of the inductor which is caused by parasitic capacitive coupling effects of the conductors. This frequency border is further reduced in the case of multi-turn and/or multilayer coils due to an intensified capacitive cross coupling of the conductors.

In [9], a figure is presented which shows the measured maximum powering range for an inductive link as a function of the frequency. It can be seen that frequencies around 10 MHz are best for fulfilling the requirements for applications aiming to remotely power devices in a distance in the centimeter to meter range. As an example, at the frequency of 10 MHz, the wavelength in free space is about 30 m, the near-field far-field border (3.7) consequently is 4.8 m and for a circular loop antenna with 50 mm radius and a copper wire with a diameter of 1 mm, the radiation resistance (3.8) of 2.4 µΩ is negligible w.r.t. the DC resistance (3.9) which is 6.9 mΩ. As mentioned before, the actual resistance may be much higher due to the skin effect.

The frequency of 10 MHz is used in [5] for powering devices over distances in excess of two meters. In [6], a slightly higher frequency of 13.56 MHz is chosen as it is part of an ISM band. If smaller antennas and shorter powering distances are used, the optimum working frequency increases due to the allowed smaller near-field far-field border, higher SRF of the coil and a lower radiation resistance.

There exist some additional factors influencing the proper working frequency of an IPT system. As will be demonstrated in section 3.3.3, high efficiencies are obtained when both coils are operated in the so-called resonance modus. This is usually achieved by adding a discrete capacitor in parallel to the coils. The requirements for the capacitors vary for different frequencies since lower frequencies demand higher capacitances. Moreover, the hardware costs and effort to generate the driving signal may change with frequency.

3.2.2 Quality Factor Definitions

An important property for the design of inductors is the so-called quality factor or simply Q-factor which relates the reactive behavior of the inductor with the occurring losses. The most general definition of the quality factor of passive devices is 2π times the stored energy per cycle divided by the energy dissipated in each cycle [52]. This definition can be applied to almost any physical system storing energy. For a passive electronic device which is operated in time-harmonic mode, the above definition can be refined as

$$Q = \frac{\text{Reactive power}}{\text{Dissipated power}}. \tag{3.12}$$

Generally spoken, the higher the quality factor the lower the losses of the device. For inductors and capacitors, the quality factor becomes infinite for ideal devices and finite for real inductors or capacitors which involve losses caused by various physical effects.

For the design of IPT systems, the quality factor of the coils is the key parameter w.r.t. the power transfer efficiency[2] because it incorporates all unwanted losses. For inductors, the definition (3.12) can be expressed more precisely by using two different definitions. When applying (3.12) to the impedance formulation of the one-port network in (2.58), the intrinsic quality factor Q_L of a coil can be written in terms of

$$Q_L = \frac{2\omega(W_\mathrm{m} - W_\mathrm{e})}{P_\mathrm{l}} = \frac{X}{R}. \tag{3.13a}$$

If the electric energy W_e is small compared to the magnetic energy W_m, the equation can be simplified by using (2.59) instead of (2.58) which results in the standard expression

$$Q_L = \frac{\omega L}{R}. \tag{3.13b}$$

In this case, the coil is modeled by a series connection of an inductance and a resistance which represents ohmic losses in the windings. Both definitions of (3.13) imply a frequency-dependent behavior of the quality factor which can be summarized as follows: Due to the linear dependence of Q_L on the frequency ω, the quality factor rises linearly with the frequency. However, this effect is attenuated at high frequencies because of the increased resistance due to the skin- and proximity-effect losses as well as the reduced internal inductance. Moreover, the capacitive cross couplings of the conductors increase the electric energy in (3.13a) at high frequencies and the quality factor is further decreased. At some specific frequency f_SRF, both electric and magnetic energy components are identically. Here, the coil operates in resonance mode which is referred to as Self-Resonant Frequency (SRF) in the following. In the resonance case, the quality-factor (3.13a) is zero. If the total reactive power is unified in the frequency-dependent inductance of (3.13b), this expression is also zero at the SRF.

The above considerations in the resonant case show that the intrinsic quality factor definition (3.13) does not adequately reflect the general quality factor definition which is basically the stored energy per cycle divided by the energy dissipated in each cycle. For this reason it is convenient to define the quality factor Q_0 of an RLC resonance circuit according to

$$Q_0 = \left.\frac{\text{Reactive power in } L \text{ or } C}{\text{Dissipated power}}\right|_{\text{at resonance}}. \tag{3.14a}$$

When this definition is rewritten with network elements, the formulation varies depending on the network topology. For a series resonance circuit, the above equation results

---
[2]This matter of fact will be detailed in section 3.3.3, especially in (3.38b).

Chapter 3 Inductive Power Transmission

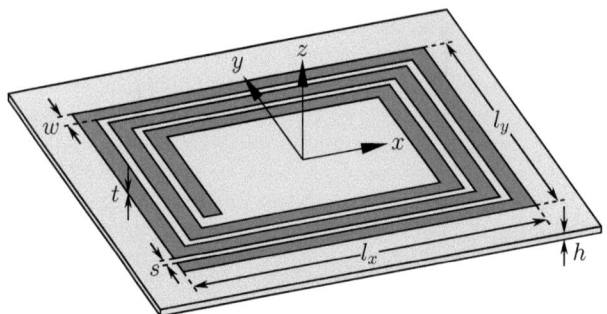

Figure 3.5: Geometrical layout of a rectangular PSC on a dielectric substrate. The coil is composed of $N_\text{turn} = 3$ turns in this example. Vias and return conductors are not shown.

in $Q_0 = (\omega_0 L)/R$ while it is $Q_0 = R/(\omega_0 L)$ for a parallel resonance circuit. In both cases, $\omega_0 = 2\pi f_0$ denotes the angular resonance frequency. It can be shown that the Q-factor of (3.14a) is identical to [17]

$$Q_0 = \frac{f_0}{\Delta f_\text{3dB}}. \tag{3.14b}$$

In (3.14b), Δf_3dB is the bandwidth of the resonance circuit, defined as the difference of two frequencies f_2 and f_1 which belong to the values where the absolute value of the resonance curve is the $1/\sqrt{2} \approx 3$ dB fraction of the maximum value.

3.2.3 Rectangular Printed Spiral Coil

In this section, the geometrical parameters of Printed Spiral Coils (PSCs) are discussed since these coil structures are used for all examples presented in this work. PSCs are a special mounting form of inductors, often being used in IPT systems. This is because PSCs are composed of planar structures which can be easily integrated into flat devices, even together with other electronic components on a single PCB. Furthermore, the reproducibility of optimized layouts is permitted since the fabrication tolerances are usually not very high. These tolerances will be focused on in section 6.2.3 on page 142 via a sensitivity analysis.

The geometrical layout of a rectangular PSC is visualized in Figure 3.5. It is characterized by a few geometrical parameters which are visualized in the drawing. The rectangular shape is often used in practical applications as it fits well into a lot of devices such as smart cards. Furthermore, it can easily be modeled via a number of straight conductor segments. Other designs with circular or partially curved shapes can be approximated by polygonal segments with piecewise straight lines. A detailed overview of different PSC shapes is given in [52].

For design and optimization purposes it is sometimes necessitated to ensure a specific part of the available coil area being filled with the conducting material. For this reason, the accumulated trace length l_c of the spiral inductor according to Figure 3.5 is introduced as[3]

$$l_c = 2N_{\text{turn}}(l_x + l_y) + 4sN_{\text{turn}} - (4N_{\text{turn}}^2 + 1)(w + s). \tag{3.15}$$

With the help of the total conductor length of (3.15), the fill factor[4] γ of the rectangular coil can be defined as

$$\gamma = \frac{l_c w}{l_x l_y}. \tag{3.16}$$

The above fill factor is the area of the coil being filled with the conductor trace normalized to the total available area of the rectangular PSC.

Moreover, it may be convenient to compute the trace width belonging to a pre-specified fill factor. This can be achieved by substituting (3.15) into (3.16) and solving for the conductor width w which results in

$$w(\gamma) = \frac{\alpha_1 - s\alpha_2}{2\alpha_2} - \sqrt{\frac{(\alpha_1 - s\alpha_2)^2}{4\alpha_2^2} - \frac{\gamma l_x l_y}{\alpha_2}}, \tag{3.17}$$

with the two abbreviations $\alpha_1 = 2N_{\text{turn}}(l_x + l_y) + 4sN_{\text{turn}}$ and $\alpha_2 = 4N_{\text{turn}}^2 + 1$.

3.3 Equivalent Circuit Representation

This section concentrates on the network description of inductively coupled antenna systems. In contrast to the preceding sections where the general behavior and functionality of a single antenna have been discussed, a system of multiple antennas is considered from now on. This is a necessity due to the fact that each IPT system generally works with at least two coupled inductors.

Traditionally, inductive applications are designed for a fixed coil arrangement and coupling behavior allowing for an optimization of the entire system. In contrast, the inductive applications regarded in this work generally come along with a possibly varying spatial positioning as can be seen from the RFID example according to Figure 3.1. In other words, it is implied that the relative position and alignment of the antennas may change during operation. This fact is of particular difficulty since unlike antenna systems with far-field coupling via electromagnetic waves, the near-field antennas can generally not be designed and optimized individually. This is because the presence of a second antenna immediately influences and changes the undisturbed field distribution of the first antenna even if the second antenna is not connected to an external circuitry.

---

[3]In (3.15), the starting point of the first outer conductor starts at a $w/2$ shift. This is convenient for attaching a feed line but not visualized in Figure 3.5.
[4]In [15] and [21], similar quantities are defined and named fill ratio ρ and fill factor φ, respectively.

Chapter 3 Inductive Power Transmission

In order to overcome the difficulty of modeling arbitrarily-positioned antenna systems, the brute force approach would be to repeatedly simulate the complete system for each variation of the geometric parameters with a 3D numerical solver and to extract the port impedances or scattering parameters for each setup. Especially for optimization purposes, this approach can be very cumbersome. This is because of the typically long simulation time needed to solve applications with a fine mesh density in the proximity of the conductors which is mandatory for skin- and proximity-effect loss modeling. In addition, the size of the problem would approximately be N-times the size of the original problem with N being the number of antennas. For this reason, a more flexible and time-saving approach being able to separate the calculation of the mutual antenna coupling from the location independent self-impedances of the antennas is sought after. The reassembly to an entire antenna system model should be maintained in a post-processing step.

The described goal can be reached by switching from the EM-field domain to the network domain. A circuit interpretation of inductively coupled antenna systems has already been motivated in the last sections. The reason for this is the dominating magnetic field of such loop antennas, which justifies to neglect the electric energy in the first instance. This, in turn, allows for interpreting the measured or simulated port impedance as a series connection of an inductance and a resistance according to (2.59). The inductance comprises the ability to store magnetic energy while the resistance incorporates ohmic losses.

A circuit representation of the near-field antennas is especially convenient for frequencies in the low MHz range and below, since almost all additional circuit elements such as required for building matching networks are realized in form of discrete network elements. Consequently, a network description of the antenna system allows an entire simulation of the system in the network domain via standard solvers like SPICE.

The circuit interpretation of the inductively coupled coil system can be achieved by the well-known network concept based on mutually coupled inductances, which is also referred to as the transformer concept. Transformers have practically been used for electrical isolation and energy and/or signal transfer for over a century [16]. When applying the transformer concept to IPT antenna systems, some differences arise because of the fact that IPT systems are coreless or at least do not hold closed cores which would prohibit an arbitrary positioning of the transmitter and receiver.

It will be shown in the following subsections that the transformer concept can be utilized for IPT systems, even under varying coupling conditions. The only prerequisite is that the antennas are represented by compact network models containing a main inductance. Afterwards, the complete system consisting of a source, two matching networks, two inductively coupled antennas and a load will be analyzed. At the end of this section, some design hints for obtaining a high overall efficiency are presented and optimum parameter settings for an RFID antenna system are derived.

3.3.1 Air Coupled Transformer Concept

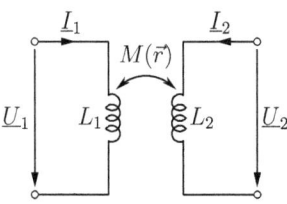

Figure 3.6: Mutually coupled inductances

The concept of mutually coupled inductances is visualized in Figure 3.6 for the case of a two-inductor arrangement. Each inductance is connected to one port of the two-port network. The mutual coupling is quantified via the mutual inductance M which is generally a function of the relative spatial positioning $\vec{r}$ of both inductors. In order to derive the two-port network expressions for this setup, use is made of Faraday's law (2.1a) in integral form. Applied to the setup as displayed in Figure 2.3, it results in

$$\underline{U}_{\text{ind}} = \int_{\partial A} \underline{\vec{E}} \cdot \mathrm{d}\vec{s} = -j\omega \int_A \underline{\vec{B}} \cdot \mathrm{d}\vec{A} = -j\omega \underline{\Psi}. \tag{3.18}$$

Comparing the above equation with (2.46) and (2.47), the two-port relation is obtained as

$$\begin{bmatrix} \underline{U}_1 \\ \underline{U}_2 \end{bmatrix} = j\omega \begin{bmatrix} L_1 & M \\ M & L_2 \end{bmatrix} \begin{bmatrix} \underline{I}_1 \\ \underline{I}_2 \end{bmatrix}. \tag{3.19}$$

As already stated in (2.45), the mutual inductance $M = L_{12} = L_{21}$ is symmetrical. Equation (3.19) clarifies that a current flow in one of the coils will induce a voltage drop in the other coil. From this relation it becomes clear that a remotely-powered receiver will also influence the primary voltage and current, respectively. In section 6.3, it will be shown for an exemplary test setup that the relation (3.19) is able to reflect the near-field coupling of the coils.

In order to apply the concept of mutually-coupled inductances (3.19) to IPT antenna systems and to simulate, design and optimize each antenna individually, two requirements have to be fulfilled. First, an adequate network extraction technique must provide the self-inductances L_1 and L_2 to describe the main diagonal of (3.19). This can be achieved by measuring or simulating the port impedance for each antenna in absence of the other antenna. The most simple case for extracting the inductance from the port impedance is to separate the impedance by real and imaginary parts via (2.59). However, especially for systems with non-negligible electric energy, e.g. for systems operating near the SRF, this approach is not feasible as the self-inductance drops down in this case and, thus, the coupling factor k of (2.48) could exceed one. In addition, if a broadband network model is desired, a simple series connection of a resistance and an inductance does not reflect the frequency-dependent behavior due to skin- and proximity-effect losses. In this case, more sophisticated parameter extraction algorithms are needed which will be focused on in the next section.

The second requirement concerns the mutual inductance computation of the two-port system (3.19). As can be seen in Figure 3.1, the mutual inductance depends on the relative position $\vec{r}$ as well as the orientation indicated by the normal vector $\vec{n}$ and the rotation β around the normal direction. Because of these spatial dependencies, an appropriate mutual inductance extraction technique should not only be accurate but also be very fast in order to allow for rapid spatial sweeps.

A method being able to approximate the mutual inductance of arbitrary coil systems is the Greenhouse method [53] in which each coil is partitioned into a set of straight filaments. Each segment of the first coil is mutually coupled with each segment of the second coil whereas the inductances are computed by evaluating (2.45) via analytical or empirical expressions as presented for a multitude of different arrangements by Grover in [54]. The total mutual inductance of (3.19) is then obtained by summing up all partial mutual inductances. In section 4.5.4, the appropriate equation will be derived, while it will be shown that the Greenhouse method can be regarded as a special case of the more general PEEC method. By using this technique, the mutual inductance can be computed in milliseconds on modern computer systems. This allows for fast spatial sweeps while maintaining a sufficiently high accuracy as will be presented in section 6.3.

It should not be concealed that the concept of mutually coupled inductances has some limitations due to the fact that it is valid under MQS assumptions only.[5] More clearly spoken, capacitive coupling effects are not accounted for in the above transformer concept. While the influence of the occurring capacitive coupling of a single coil can be regarded in its equivalent circuit representation (s. section 3.3.2), the capacitive cross coupling of wires belonging to different inductors is not incorporated in this model. This can cause errors especially if both coils are located in close proximity of each other [55]. To overcome this limitation, both coils can be simulated in an entirely numerical model. Alternatively, the model in Figure 3.6 could be extended by one or more lumped capacitances which would be connected with one terminal on the primary side and the other terminal on the secondary side, respectively. In this case, the capacitance values would depend on the spatial arrangement of the coils.

The transformer concept can easily be extended to inductor systems with more than two coils. In this case, the main diagonal matrix elements represent the self-inductances while each off-diagonal element characterizes a mutual inductance of the corresponding pair of coils.

3.3.2 Antenna Impedance Macromodeling

This section will systematically lead to different equivalent circuit models which all approximate the frequency-dependent behavior of a single coil. The presented methods work for any kind of linear, passive component which is dominated by the magnetic energy at frequencies below the first SRF. Besides the presentation of different narrow-band models, a broadband model is focused on which is valid from DC up to the first

---

[5]The reason for being valid under MQS assumptions only is the use of the inductance expressions based on the magnetic flux of (2.46) and (2.47).

3.3 Equivalent Circuit Representation

SRF. For frequencies beyond the SRF, the device acts alternatingly in a capacitive or inductive way [52]. This behavior is not captured by the proposed network models.

The basic network component of all models is a main inductance which is needed to describe the transformer concept of the last section. Depending on the desired accuracy and acceptable effort, different additional elements are added in order to account for the physical relevant behavior. All extracted parameters are based on lumped elements allowing for simulations in time and frequency domain with standard circuit solvers such as SPICE.

Before discussing the different network models in detail, some state-of-the-art information about coil macromodeling is presented. The description of coils by means of compact network models has been discussed in many publications. A very often used model is the so-called "nine-element" π-model, e.g. [21, 56], which has basically been developed for planar inductors applied on lossy substrates on top of a ground plane. Some contributions extend this model in order to account for frequency-dependent loss mechanisms. This can be achieved by either introducing a set of small transformer loops which are coupled with the main inductance [57, 58] or by replacing the series resistance by a ladder circuit consisting of frequency independent RL elements [59]. As this approach is also being used in this work, a detailed description will be presented in section 3.3.2.2. A third macromodel which accounts for the frequency-dependent losses by partitioning the standard π-model into two coupled halves is presented in [60].

In order to extract the circuit parameters of any of the models, it is implied throughout this section that an MQS *and* an LQS simulation tool are available. Alternatively to LQS, a full-wave solver can be applied assuming that the radiation is of negligible consequence. Furthermore, both simulation models should be identical in terms of mesh density or solver accuracy since the difference of both port impedances is evaluated for the parameter extraction technique. Thus, different solver setups could lead to additional sources of error if small impedance differences have to be evaluated.

As mentioned before, the numerical approach used in this work is the PEEC method, though any other numerical EM method with the described prerequisites would work as well. On the other hand, it is obvious that the algorithms do not work with measured impedance data as it is not possible to separate the measured impedance into full-wave and MQS components. In this case curve fitting techniques like vector fitting [61] can be applied which approximate the frequency-dependent behavior of the impedance without any a priori knowledge of the structures. However, as these fitting algorithms are of mathematical nature, they do not necessarily provide the main inductance which is mandatory for the transformer concept.

In contrast to that, the methods presented here aim to determine the parameters of a predefined network topology aspiring to approximate the main physical behavior of the coil as closely as possible in order to be consistent with the transformer concept. The number of network elements is kept as small as possible in order to account for the

Chapter 3 Inductive Power Transmission

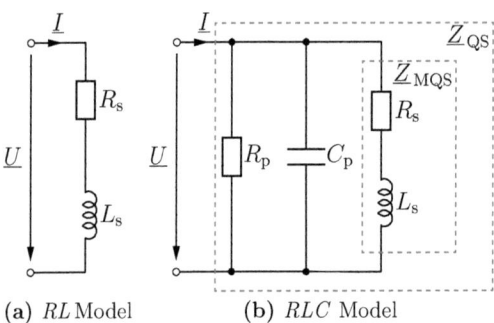

(a) *RL* Model (b) *RLC* Model

Figure 3.7: Two different macromodels of a single coil. (a) Separation of the impedance by real and imaginary parts. (b) Resonance circuit obtained by two different simulation models.

frequency-dependent behavior by a small number of well-matched parameters. Results for an exemplary PSC are presented in section 6.2.3 on page 138.

3.3.2.1 Narrowband Model

In the setup presented in Figure 3.7, two different network topologies for approximating the frequency-dependent port impedance of a single coil are visualized. Each of the models will be focused on in the following two paragraphs.

RL model The most simple network description of a single coil is shown in Figure 3.7a, in which the impedance is specified by

$$\underline{Z} = R_\text{s} + j\omega L_\text{s}, \tag{3.20}$$

with R_s being the serial resistance and L_s the serial inductance, respectively. Due to the fact that in MQS systems no electric energy exists,[6] the expression (3.20) is always valid when using an MQS solver. This is because in the MQS formulation, the partitioning of the impedance via (2.59) into a resistance and an inductance is always possible. Thus, the parameters are extracted by separating the simulated port impedance $\underline{Z}_\text{MQS}$ at the single frequency ω into real and imaginary parts according to

$$R_\text{s} = \text{Re}\{\underline{Z}_\text{MQS}\}, \qquad L_\text{s} = \text{Im}\{\underline{Z}_\text{MQS}\}/\omega. \tag{3.21}$$

This model does not take into account the frequency-dependent behavior due to the eddy currents but rather uses the computed values at the chosen frequency. The above partitioning allows for directly computing the intrinsic quality factor Q_L of the coil via (3.13b).

---

[6]This can be verified by rewriting Poynting's theorem (2.35) under MQS assumptions. Due to the neglect of the displacement current, the electric energy does not appear in the equation.

3.3 Equivalent Circuit Representation

Because of the neglect of the electric energy in the MQS formulation, $\underline{Z}_{\mathrm{MQS}}$ can also be used to approximate the resonance quality factor Q_0 of (3.14a) if the SRF is known.

Formally, the parameter extraction of (3.20) can also be realized by using an LQS solver, a full-wave solver or measurements. In all mentioned cases, the partitioning of the impedance into real and imaginary parts can be realized according to (3.21). However, the part of the electric energy from (2.58) which is also included in L_s, affects the inductance as well as the intrinsic quality factor Q_L to be identically zero at the SRF. Beyond this frequency, the network model according to Figure 3.7a is unphysical as the inductance is negative in this case.

RLC model A network representation capturing the behavior in the frequency range near the SRF more precisely than the RL model is the so-called RLC model[7] as presented in Figure 3.7b. As depicted by the dashed lines in the figure, two different simulations $\underline{Z}_\mathrm{QS}$ and $\underline{Z}_\mathrm{MQS}$ need to be performed at the desired frequency. Introducing the difference in admittance and impedance of the two simulated impedances as

$$\underline{Y}_\mathrm{D} = \frac{1}{\underline{Z}_\mathrm{QS}} - \frac{1}{\underline{Z}_\mathrm{MQS}}, \qquad \underline{Z}_\mathrm{D} = \frac{1}{\underline{Y}_\mathrm{D}}, \qquad (3.22\mathrm{a})$$

allows for computing the parameters presented in Figure 3.7b according to

$$R_\mathrm{s} = \mathrm{Re}\{\underline{Z}_\mathrm{MQS}\}, \qquad L_\mathrm{s} = \mathrm{Im}\{\underline{Z}_\mathrm{MQS}\}/\omega, \qquad (3.22\mathrm{b})$$
$$R_\mathrm{p} = 1/\mathrm{Re}\{\underline{Y}_\mathrm{D}\}, \qquad C_\mathrm{p} = \mathrm{Im}\{\underline{Y}_\mathrm{D}\}/\omega. \qquad (3.22\mathrm{c})$$

It is seen that R_s and L_s of (3.22b) are defined identically to (3.21) of the RL model. The parallel capacitance C_p accounts for the electric part of the energy while the parallel resistance R_p is included in the model for the following two reasons: First, R_p is required to uniquely map the two real parts and two imaginary parts of the simulated impedances to the four parameters of R_s, L_s, R_p, and C_p. Second, the resistance allows the inclusion of dielectric losses which cannot be modeled in the MQS impedance.

If no dielectric losses are modeled, R_p is typically in the range of several MΩ and can be neglected without reasonable errors. The network topology displayed in Figure 3.7b provides the following expression for the SRF

$$\omega_\mathrm{SRF} = \sqrt{\frac{1}{L_\mathrm{s} C_\mathrm{p}} - \frac{R_\mathrm{s}^2}{L_\mathrm{s}^2}} \approx \frac{1}{\sqrt{L_\mathrm{s} C_\mathrm{p}}}, \qquad (3.23)$$

which is obtained by solving the total admittance according to Figure 3.7b for zero imaginary part. It should be mentioned that the SRF obtained by an LQS solver can

---
[7]The RLC model is structurally similar to the "nine-element" π-model of [21] without including the circuit elements accounting for the substrate influence.

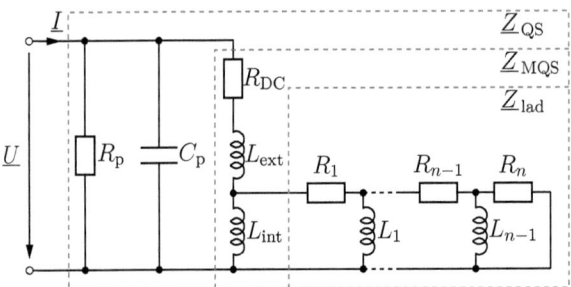

Figure 3.8: Broadband macro-model of a PSC. The resonance model is chosen equivalently to the setup shown in Figure 3.7b with L_ext being the main inductance. Frequency-dependent losses are accounted for by the ladder model $\underline{Z}_\text{lad}$.

deviate from the SRF obtained by a full-wave solver because in the LQS approximation only a part of the electric energy is accounted for.[8]

According to the RL model, the RLC model presented in this paragraph does not take into account the frequency-dependent behavior of the resistance and inductance. However, it is able to capture the resonance behavior quite well. This is due to the fact that the parameter change of R_s, L_s, R_p, and C_p with frequency is relatively small. As a consequence of the presented circuit topology visualized in Figure 3.7b, both definitions of the quality factors Q_L and Q_0 are available since the magnetic and electric energies are separated into the lumped elements L_s and C_p.

3.3.2.2 Broadband Model

In some applications it might be necessary to approximate the *full spectrum* of the coil impedance from DC up to the first SRF by an equivalent circuit representation based on frequency independent lumped elements. Hereby, circuit simulations in time and frequency domain are enabled and the correct DC behavior is granted. A network topology which extends the models of the preceding paragraphs is shown in Figure 3.8.[9] The main inductance[10] is represented by the external inductance L_ext whereas the DC losses are accounted for by R_DC. The internal inductance at DC is represented by L_int. As before, the parasitic capacitance is modeled by C_p (together with R_p) whereas the eddy currents inside the conductors are now accounted for by a ladder model characterized by $\underline{Z}_\text{lad}$. This approach allows for modeling skin- and proximity-effect losses inside the conductors as well as frequency-dependent inductance. Because of the specific ladder-model topology presented in Figure 3.8 with the resistances R_1 to R_n and inductances L_1 to

---

[8]This can be pointed out when repeating the derivation of the Poynting's theorem in the LQS formulation.
[9]The network visualized in Figure 3.8 is similar to the variant presented in [59].
[10]In this case, it is the external inductance L_ext instead of L_s to which possibly other coils are coupled with. This is no limitation because the inductance definition via magnetic fluxes is valid only if $L_\text{ext} \approx L_\text{tot}$.

3.3 Equivalent Circuit Representation

L_{n-1}, the total DC inductance L_{DC} can be read directly from the figure as

$$L_{DC} = L_{ext} + L_{int}. \qquad (3.24)$$

This can be verified by the fact that no current flows through the ladder circuit $\underline{Z}_{lad}$ at the DC limit.

The parameter values of the circuit elements according to Figure 3.8 are obtained by a fitting technique which requires the analysis of the LQS and MQS models at certain frequency points. The parameter extraction technique is partitioned into three main steps according to [55]:

First step In the first parameter extraction step, the SRF f_{SRF} of the inductor is estimated by an iterative process. The knowledge of f_{SRF} determines the valid frequency range whereas the algorithm provides the parameters C_p and R_p of the network model as a by-product. As already mentioned before, the simulated impedance $\underline{Z}_{MQS}$ involves the frequency-dependent inductive and resistive effects. The impedance $\underline{Z}_{QS}$ in addition includes the capacitive behavior.

By defining the network topology in this first step according to Figure 3.7b, the iterative process starts with simulating both the LQS and MQS system at an arbitrarily chosen medium frequency f_1. Equations (3.22) are then used to determine the parameters R_s, L_s, R_p, and C_p which in turn provide an approximation for the SRF in (3.23). A new LQS and MQS system analysis can be performed at the estimated SRF. Repeating the preceding steps at the updated frequency improve the estimation of the parameters. Since the change of the parameters R_s, L_s, R_p, and C_p with frequency is relatively small, this iterative process converges fast. For typical arrangements as presented in [55], only two iteration steps are required to get a relative error in the SRF below 1%.

At the end of the first fitting step, two MQS and two LQS simulations have been performed. The parameters R_p and C_p are determined using the values of the last iteration step near the SRF of the coil.[11] Contrary, the parameters R_s and L_s are discarded because the RL behavior will be modeled more detailed in the second and third fitting step. Nevertheless, the two simulated MQS impedance values are stored in order to be reused for obtaining the frequency-dependent skin and proximity losses in the third fitting step.

By utilizing the parameter values in the described manner, it is obvious that the magnetic part of the model is paid more attention. This is justified by the facts that on the one side, the magnetic energy always dominates in inductive applications and on the other side, it is sufficient to model the capacitive behavior near the SRF by using a single capacitance only.

---

[11] Generally, the dielectric losses characterized by R_p follow a $1/f$-characteristic, e.g. [62]. This can either be ensured by using a frequency-dependent resistance as $R_{p,freq}(f) = R_p f_{SRF}/f$ or by introducing a ladder-type circuit model as presented in [62] if lumped network elements are favored.

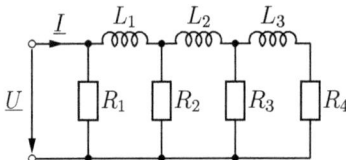

Figure 3.9: Equivalent ladder circuit for modeling the skin effect according to [65] consisting of four stacked R-elements and three L-elements.

Second step The second step is dedicated to the computation of the parameter values R_{DC}, L_{ext}, and L_{int} as specified in Figure 3.8. For this purpose, only two additional MQS analyses are required. The first MQS simulation is performed at a very high frequency where the skin penetration and the magnetic energy inside the conductors, respectively, are negligible. Thus, the total inductance is almost equal to the external inductance

$$L_{\text{ext}} = \text{Im}\{\underline{Z}_{\text{MQS}}\}/\omega. \tag{3.25}$$

This fact can also be verified by referring to Figure 3.8, in which the total MQS current bypasses the inductances L_{int} and L_1 to L_{n-1} at frequencies approaching infinity. Consequently, the only reactive component contributing to $\underline{Z}_{\text{MQS}}$ is L_{ext} and the expression (3.25) is justified.

In order to extract the internal inductance in the DC case, a further MQS simulation is evaluated at a very low frequency. Here, the current density inside the conductors is almost homogenous, yielding

$$L_{\text{int}} = \text{Im}\{\underline{Z}_{\text{MQS}}\}/\omega - L_{\text{ext}}, \qquad R_{\text{DC}} = \text{Re}\{\underline{Z}_{\text{MQS}}\}. \tag{3.26}$$

The DC resistance R_{DC} is obtained by this analysis simultaneously.

Third Step The last fitting step is focused on the modeling of the frequency-dependent behavior of the MQS impedance due to skin and proximity effects which generally imply an increasing resistance and a decreasing inductance with rising frequency. The authors in [63, 64] have physically motivated to model this behavior by a number of stacked lumped RL elements which is referred to as a ladder model. This technique has been concretized by Kim and Neikrik [65] to a circuit model consisting of four resistances and three inductances as presented in Figure 3.9. Additionally, a methodology to extract the parameter values by introducing a constant resistance ratio factor RR with $R_i/R_{i+1} = RR$ and $L_i/L_{i+1} = LL$, respectively, is presented in the reference. In [59], the network topology displayed in Figure 3.9 has been used to extend the standard "nine-element" π-model to 15 lumped network elements. The benefit of this model is the ability to account for eddy-current losses in the conductors.

The limitation of the constant resistance and inductance ratios in each stage of the ladder circuit has been overcome by Görisch [66] who presents an approach to determine the parameter values independently of each other. Furthermore, the number of network

3.3 Equivalent Circuit Representation

elements is no longer restricted to seven in total. Although this approach has been used by Görisch to model the skin impedance of a single conductor, it can be extended to account for a spiral coil with skin and proximity effects assuming that a general fitting algorithm is applied to extract the parameter values.

When comparing the ladder models presented in Figure 3.8 and Figure 3.9, a slightly different topology of both models can be observed. In the variant displayed in Figure 3.8 which is used in this work, the internal DC inductance has been dragged out of the ladder circuit explicitly since it has already been determined by the second step of this fitting procedure. Thus, the correct DC behavior is ensured even if the fitting algorithm of the ladder-circuit parameters would give wrong parameter values. This is obvious because the internal inductance shorts the ladder circuit at DC.

In order to extract the parameter values of the ladder-model topology in Figure 3.8, the elements R_i and L_i are optimized independently of each other by simulating the MQS impedance at a small number of logarithmic spaced frequency points ω_i. As mentioned before, the stored values from the first fitting step are reused. The elements R_{DC}, L_{int}, and L_{ext} which have already been determined are subtracted from $\underline{Z}_{\text{MQS}}$ leading to

$$\underline{Z}_{\text{lad}}(j\omega_i) = \left(\frac{1}{\underline{Z}_{\text{MQS}}(j\omega_i) - R_{\text{DC}} - j\omega_i L_{\text{ext}}} - \frac{1}{j\omega_i L_{\text{int}}} \right)^{-1}. \qquad (3.27)$$

This ladder impedance $\underline{Z}_{\text{lad}}$ is approximated by a fitting algorithm based on [67] which gives a rational polynomial in $j\omega$ as

$$\underline{Z}_{\text{lad}}(j\omega) = \frac{b_0 + b_1 j\omega + \ldots + b_{n-2}(j\omega)^{n-2} + b_{n-1}(j\omega)^{n-1}}{a_0 + a_1 j\omega + \ldots + a_{n-2}(j\omega)^{n-2} + 1(j\omega)^{n-1}}. \qquad (3.28)$$

In order to extract the parameters of the network elements from the coefficients a_i and b_i, different methods can be applied. While in [57, 58], the coefficients are converted to a number of small transformer loops being coupled with the main inductance, the method presented here converts the coefficients to the parameters of the ladder model. According to [66], this can be achieved by performing a continued-fraction decomposition of (3.28) as presented in [68]. This allows the computation of the parameter values in a straightforward manner resulting in

$$\underline{Z}_{\text{lad}}(j\omega) = R_1 + \cfrac{1}{\cfrac{1}{j\omega L_1} + \cfrac{1}{\ldots + R_{n-1} + \cfrac{1}{\cfrac{1}{j\omega L_{n-1}} + \cfrac{1}{R_n}}}}. \qquad (3.29)$$

In practice, one has to ensure that the parameter values are chosen from a physically valid range. Case studies in [55] have shown that approximation orders of four to five are needed for sufficient accuracy.

Chapter 3 Inductive Power Transmission

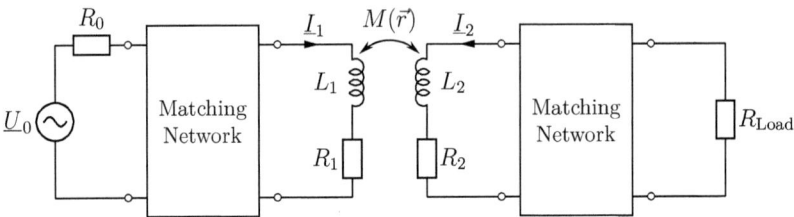

Figure 3.10: Equivalent circuit of a wireless energy transfer system consisting of a source, two matching networks, the mutually coupled inductors as well as a resistive load. The presented network covers all coil models from Figure 3.7 and Figure 3.8. This is due to the facts that the capacitive effects are integrated in the matching networks and R_1, R_2, L_1, and L_2 may be frequency dependent.

To sum up, a broadband model consisting of about 14 lumped RLC elements allow for an accurate frequency-dependent description of the port impedance of a PSC. Results of the fitting algorithm will be presented in section 6.2.3 on page 138.

3.3.3 System Design

In this section, the complete IPT system consisting of a source, two inductively coupled antennas as well as a load R_{Load} as depicted in Figure 3.10 are considered. The source is characterized by a voltage source $\underline{U}_0$ together with the internal resistance R_0. A subsequent Matching Network (MN) is required in order to be able to maximize the overall power by means of impedance matching.

Generally, an ideal matching network can be used to convert a given output impedance into an arbitrary input impedance at a single frequency. The matching network at the transmitting or primary circuit is connected in between the source and the transmitting coil consisting of the inductance L_1 and the resistance R_1. These elements may either be the frequency independent parameters according to Figure 3.7 at a specified working frequency or, alternatively, may be replaced by the MQS model visualized in Figure 3.8 if a broadband analysis is required.

According to the last section, the primary coil is inductively coupled with the secondary coil which is composed of the inductance L_2 as well as the resistance R_2 with both values being defined equivalently to the first coil. The capacitive behavior of both coils which exists in the models displayed in Figure 3.7b and Figure 3.8, can be included in the matching networks. Consequently, it does not need to be considered explicitly in the network model as presented in Figure 3.10.[12]

---

[12] In section 3.3.3.2, a concretized matching network topology for regarding the capacitive effects is presented.

3.3 Equivalent Circuit Representation

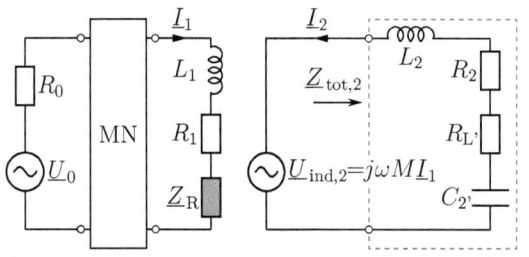
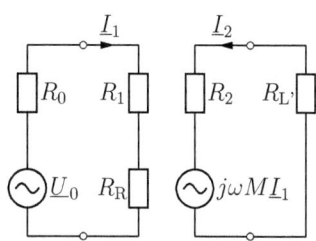

(a) Equivalent system with the most simple matching network consisting of a single capacitance

(b) Resonance model of ohmic resistances only

Figure 3.11: Equivalent circuit representation of the IPT system. (a) The mutual coupling is replaced by the induced voltage at the secondary circuit and by the reflected impedance on the primary circuit, respectively. In the resonance case (b), all reactive elements compensate each other. In this case, the whole system consists of pure ohmic components.

Subsequent to the receiving coil, a second matching network is appended to the system which is used to compensate the reactive power of the secondary coil, to influence the overall efficiency and to obtain the maximum power as will be examined in section 3.3.3.1.

The system specified in Figure 3.10 is completed by the load which is represented by a single resistance in the following. Usually, the load is composed of a rectifier circuit to convert the AC into a DC voltage which subsequently can be used to supply a digital circuit and/or, for instance, to charge a battery (cf. [9, 69]). The load circuit generally involves an additional capacitive part. For keeping the expressions as simple as possible, this part is accounted for in this section as being a part of the second matching network.

In the following, the necessary equations to describe the entire system behavior are deduced. For this reason, the IPT system is considered in a slightly modified way as sketched in Figure 3.11. First, the secondary matching network has been set to the most simple case consisting of a series capacitance $C_{2'}$ and a load resistance $R_{L'}$.[13] Second, the coupling via the mutual inductance has been replaced by a voltage source on the secondary circuit and an impedance on the primary side, respectively. This can be obtained by expressing the induced voltages of (3.19) as a function of the driving current $\underline{I}_1$ according to

$$\underline{U}_{\text{ind},2} = j\omega M \underline{I}_1, \tag{3.30a}$$

$$\underline{U}_{\text{ind},1} = j\omega M \underline{I}_2 = -j\omega M \frac{\underline{U}_{\text{ind},2}}{\underline{Z}_{\text{tot},2}} = \frac{\omega^2 M^2}{\underline{Z}_{\text{tot},2}} \underline{I}_1 = \underline{Z}_R \underline{I}_1, \tag{3.30b}$$

---

[13] The load resistance is named $R_{L'}$ in this simplified model instead of R_{Load} in order to avoid inconsistencies when defining the real load resistance in section 3.3.3.2.

Chapter 3 Inductive Power Transmission

whereas the conversions can be verified by Figure 3.11a. The impedance $\underline{Z}_{\text{tot},2}$ characterizes the total impedance of the receiver and can be defined independently of the specific matching network topology. In the last equality of (3.30b), the reflected impedance

$$\underline{Z}_{\text{R}} = \frac{\omega^2 M^2}{\underline{Z}_{\text{tot},2}} \tag{3.31}$$

has been introduced which allows for interpreting the receiver as a complex load impedance at the primary circuit.

The effective power delivered to the reflected impedance should be maximized because this is exactly the power which can be used for supplying the receiver. For a given current $\underline{I}_1$, the transferred effective power at the specific frequency ω_0 increases linearly with $\text{Re}\{\underline{Z}_{\text{R}}\}$ because of $\text{Re}\{\underline{P}\} = |\underline{I}_1|^2 \, \text{Re}\{\underline{Z}_{\text{R}}\}$.[14] Due to the real nominator of (3.31), the maximum effective power is transferred when the imaginary part of the denominator $\text{Im}\{\underline{Z}_{\text{tot},2}\} = 0$ vanishes.[15] In other words, the transferred effective power is maximized if the receiver is operated in resonance mode where all reactive elements compensate each other. In some contributions, a distinction between inductive coupling and resonant inductive coupling can be found. Since the underlying physical principles are identical for both cases, such a distinction is not being used in this work. Instead, the resonance condition is included in the impedance matching.

In the most simple case which is considered in Figure 3.11a, the resonance condition is ensured by the matching network series capacitance $C_{2'}$. Thus, the standard expression for the resonance case is obtained as

$$\omega_0 = \frac{1}{\sqrt{L_2 C_{2'}}}, \qquad C_{2'} = \frac{1}{\omega_0^2 L_2}. \tag{3.32}$$

When substituting the capacitance $C_{2'}$ of (3.32) into the expression for $\underline{Z}_{\text{tot},2}$ presented in Figure 3.11a, the following conversion can be made

$$\underline{Z}_{\text{tot},2} = R_2 + R_{\text{L}'} + j\omega L_2 + \frac{1}{j\omega C_{2'}} = \omega_0 L_2 \left[\underbrace{\left(\frac{R_2 + R_{\text{L}'}}{\omega_0 L_2}\right)}_{d_2} + j \underbrace{\left(\frac{\omega}{\omega_0} - \frac{\omega_0}{\omega}\right)}_{\nu} \right], \tag{3.33}$$

with $d_2 = 1/Q_{02}$ being the dissipation factor defined as the reciprocal value of the resonance quality factor from (3.14a). In addition, the frequency deviation $\nu = \omega/\omega_0 - \omega_0/\omega$ has been introduced equivalently to [12, 70]. The frequency deviation can be regarded as a normalized frequency which is zero at the resonance frequency. By means of the

---
[14]As before, RMS values are assumed whenever power quantities are involved.
[15]It is assumed that $\text{Re}\{\underline{Z}_{\text{tot},2}\}$ stays constant.

3.3 Equivalent Circuit Representation

above substitutions in combination with the coupling factor definition of (2.48), the reflected impedance can be expressed in a more compact form as

$$\underline{Z}_\mathrm{R} = \frac{\omega^2}{\omega_0^2} \frac{k^2 \omega_0 L_1}{d_2 + j\nu}. \tag{3.34a}$$

As already expected before, in case of the resonance frequency, $\omega = \omega_0$ and $\nu = 0$, the reflected impedance becomes a real value

$$R_\mathrm{R} = \omega_0 k^2 L_1 Q_{02} = \frac{\omega_0^2 M^2}{R_2 + R_\mathrm{L'}}, \tag{3.34b}$$

which is therefore also named reflected resistance R_R. Equations (3.34) can also be derived for a different matching network topology as shown in [70]. The concept of the reflected impedance (3.34a) will be used in section 6.2.3 on page 139 for a contactless measurement of the SRF and associated quality factor of a multi-turn spiral coil.

3.3.3.1 Efficiency Maximization and Field-Emission Minimization

It will be examined in this section how the IPT system can be optimized in terms of efficiency and field emission, assuming that arbitrary matching networks consisting of ideal reactive elements are available.

If the system is operated in resonance mode in which all reactive components cancel each other, a network of pure ohmic resistances can be set up according to Figure 3.11b. Although this model is valid at the resonance frequency ω_0 only, it works independently of the specific realization of the matching network. In the specified figure, the resistance $R_\mathrm{L'}$ is defined as being the load resistance transformed by an arbitrary matching network. If the matching network is realized as presented in Figure 3.11a, the transformed load is identical to the real load. However, this is not necessarily the case for different matching networks.

In general, a matching network which consists of ideal L- and C-elements, is able to transform any load impedance into an arbitrary resistance [48]. As a consequence, $R_\mathrm{L'}$ can be regarded as a design parameter. Normalizing this value by the loss resistance of the receiving coil allows the introduction of a scaling factor α according to [6] as[16]

$$\alpha = \frac{R_\mathrm{L'}}{R_2}. \tag{3.35}$$

The scaling factor allows for specifically partitioning the total transferred power into the power transferred to the load on the one side and the unwanted power dissipated

---

[16] In [6], the scaling factor α is referred to as dimensioning factor.

by the ohmic losses of the conductors on the other side. As a lower bound, $\alpha_{\min} = 1$ can be concluded. In this case, the power is partitioned to $R_{L'}$ and R_2 in equal parts.

The system behavior can be described more precisely by introducing different efficiencies η according to the general relation

$$\text{Efficiency} = \frac{\text{Useful power output}}{\text{Total power input}}. \quad (3.36)$$

Following the procedure of [6], two different efficiencies for the secondary and primary parts are introduced as

$$\eta_2 = \frac{|I_2|^2 R_{L'}}{|I_2|^2 (R_2 + R_{L'})} = \frac{\alpha}{1+\alpha}, \quad (3.37a)$$

$$\eta_1 = \frac{|I_1|^2 R_R}{|I_1|^2 (R_1 + R_R)} = \frac{1}{1 + \dfrac{R_1 R_2 (1+\alpha)}{\omega_0^2 M^2}}, \quad (3.37b)$$

in which the relations (3.34b) and (3.35) have been regarded for obtaining the above expressions. When replacing the mutual inductance by the coupling factor definition of (2.48) and using the intrinsic quality factor definitions of (3.13b), the total efficiency η can be expressed as

$$\eta = \eta_1 \eta_2 = \frac{\alpha}{1 + \alpha + \dfrac{(1+\alpha)^2}{k^2 Q_{L1} Q_{L2}}}. \quad (3.37c)$$

It is shown in [6] that the total efficiency is maximized at the optimum scaling factor

$$\alpha_{\text{opt}} = \sqrt{1 + k^2 Q_{L1} Q_{L2}}, \quad (3.38a)$$

which confirms the lower bound $\alpha_{\min} = 1$ from above.[17] If the optimum scaling factor is substituted in (3.37c), the theoretically maximum available efficiency can be expressed after some algebraic conversions as

$$\eta_{\max} = \frac{k^2 Q_{L1} Q_{L2}}{\left(1 + \sqrt{1 + k^2 Q_{L1} Q_{L2}}\right)^2}. \quad (3.38b)$$

From (3.38b) it is obvious that for increasing quality factors (decreasing losses), the efficiency approaches one. At the other extreme, zero coupling results in an efficiency of zero as expected. As an example, if two identical coils with a Q-factor of 200 each are positioned in such a way that a coupling factor of 1 % is ensued, the theoretically

---

[17]The same optimum scaling factor can be derived by using the *coupled-mode theory* [5].

3.3 Equivalent Circuit Representation

maximum overall efficiency amounts 38 % while the optimum scaling factor is about 2.2. If the coupling factor is increased to 5 %, e.g. due to a spatial shift, the maximum efficiency increases to 82 % while the optimum scaling factor is raised to approximately 10.

This examples illustrates the need for an adaptive matching network which is able to change the scaling factor in order to account for varying coupling conditions in an optimum manner. Otherwise, a non-optimum efficiency has to be accepted for coupling deviations w.r.t. the nominal coupling factor.

A second important parameter for designing an IPT system is the electromagnetic field emission which should be minimized in order to reduce the interactions with other electronic devices and human beings. Due to the dominating magnetic fields occurring in IPT systems, it is aimed to minimize the overall magnetic energy emitted by the coil system. When transferring this statement to an optimum scaling factor, the following result is obtained [6]

$$\alpha_{\text{opt},W_\text{m}} = \sqrt{1 + k^2 Q_{L2}^2}, \qquad (3.39)$$

which is similar to (3.38a). When both optimum scaling factors (3.38a) and (3.39) are compared, it turns out that the difference is given by the missing quality factor of the primary coil in (3.39). This can be explained by the fact that the losses characterized by R_1 do not influence the magnetic energy of the primary coil. If, in addition, both coils are identical or at least share the same quality factor, both optimum scaling factors for maximum efficiency and minimum total magnetic energy coincide.

For any of the two goals, i.e. maximum efficiency and minimum field emission, it has been pointed out by the above equations that it is essential to accurately model the loss resistances of the coils due to skin and proximity effects in order to allow for a precise forecast of the quality factors. This is of particular importance when choosing an appropriate numerical method for the simulation of IPT antenna systems.

The key results of this section can be used to summarize the steps needed for designing an efficiency optimized IPT system in multiple subsequent steps. First, the range of the coupling factor has to be determined. The coupling factor is mainly influenced by the size of both coils and their relative arrangement,[18] whereas other parameters such as the number of turns or conductor cross sections influence the coupling behavior only marginally. Second, the two coils have to be optimized w.r.t. the quality factors Q_{L1}, respective Q_{L2}, in order to maximize the efficiency in (3.38b). For a specified nominal coupling factor and the quality factors from above, an optimum scaling factor α_opt can be computed according to (3.38a). The matching network of the receiver is subsequently designed in such a way as to transform the given load resistance to $R_{L'} = \alpha_\text{opt} R_2$ while compensating the L_2-reactance by its complex conjugate value. If the system is operated under varying coupling conditions, an adaptive matching network must be applied which

---

[18]Parameter studies for different turn configurations are presented in [14].

Chapter 3 Inductive Power Transmission

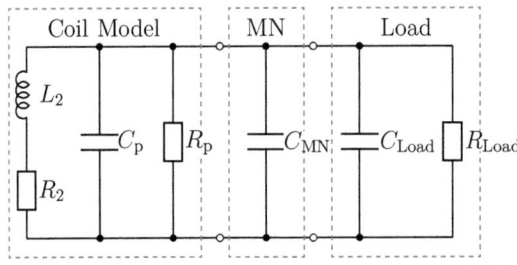
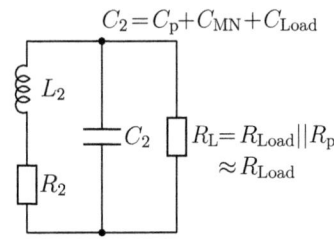

(a) Equivalent circuit model consisting of the coil, the matching network and the load

(b) Combining parallel capacitances and resistances

Figure 3.12: Typical equivalent circuit representation of the receiver. (a) The coil is modeled via the RLC circuit displayed in Figure 3.7b. The matching network consists of a single capacitance only while the load is modeled by a parallel connection of a resistance and a capacitance. (b) The entire model can be simplified by combining the parallel circuit elements.

is able to change the scaling factor according to (3.38a). Alternatively, a non-optimum system behavior has to be accepted.

3.3.3.2 Example: Design of an RFID Transponder Label

In this section, the results from the previous section are transferred to the design of a typical RFID transponder operating at the frequency of $f_0 = 13.56$ MHz (s. Figure 3.1). The PSC is modeled via the RLC circuit according to Figure 3.7b whereas the transponder Integrated Circuit (IC) is represented by a parallel circuit consisting of the load resistance as well as a capacitance.[19] In low-cost RFID systems, the matching network at the transponder must be as simple and robust as possible which is ensured by a single capacitor[20] connected in parallel[21] to the terminals of the coil.

The described parts of the transponder label are presented in Figure 3.12a, whereas the parallel network elements can be combined to C_2 and R_L as shown in Figure 3.12b. It is obvious that the parasitic capacitance C_p of the PSC has to be considered as it influences C_2 and therewith the resonance frequency of the entire system. Since a numerical modeling of the coil under MQS assumptions only is not able to model the parasitic capacitance, such an approach would unavoidably lead to a wrong modeling of the resonance frequency.

---

[19] The capacitance mainly accounts for the internal behavior of the transponder IC, e. g. due to the diode capacitance of the rectifier circuit or a tuning capacitor [9, 69].

[20] It is assumed that the internal capacitance of the IC is not large enough to fulfil the resonance condition.

[21] If the capacitor would be connected in series, the scaling factor α of (3.35) would be too high due to the large load resistance which is typically in the range of several kΩ w. r. t. the small resistance R_2 of the conducting traces which usually does not exceed a few Ω for good conductors.

3.3 Equivalent Circuit Representation

In order to agree with the outcomes presented in Figure 3.11, the parameters of the parallel RC circuit of Figure 3.12b are converted to the series connection displayed in Figure 3.11a according to the relation

$$C_{2'}(j\omega) = C_2 + \frac{1}{\omega^2 C_2 R_\mathrm{L}{}^2} \approx C_2, \qquad (3.40\mathrm{a})$$

$$R_{\mathrm{L}'}(j\omega) = \frac{R_\mathrm{L}}{1+\omega^2 C_2^2 R_\mathrm{L}{}^2} \approx \frac{1}{\omega^2 C_2^2 R_\mathrm{L}}. \qquad (3.40\mathrm{b})$$

The above approximations are valid if $\omega^2 C_2^2 R_\mathrm{L}{}^2 \gg 1$ which is often valid for typical arrangements [71]. For simplicity reasons, only the approximated versions will be used in the following. Due to the fact that the capacitance in (3.40a) stays approximately the same as in the series circuit, the resonance condition (3.32) holds and the capacitance is chosen as $C_2 = 1/(\omega_0^2 L_2)$. From this, the expressions for the transformed load impedance and the scaling factor at the working frequency ω_0 are determined to

$$R_{\mathrm{L}'} = \frac{\omega_0^2 L_2^2}{R_\mathrm{L}}, \qquad (3.41)$$

$$\alpha = \frac{\omega_0^2 L_2^2}{R_2 R_\mathrm{L}} = Q_{L2} \frac{\omega_0 L_2}{R_\mathrm{L}}. \qquad (3.42)$$

By using this specific matching network, the scaling factor α is a function of the inductance as well as the intrinsic quality factor. Thus, it cannot be chosen arbitrarily although demanded for maximum efficiency of (3.38b).

The question is whether this network can still be used to obtain efficiencies close to the maximum value by choosing the scaling factor according to (3.38a). It will be seen in section 6.2.3 on page 132, where different design parameters of the rectangular PSC according to Figure 3.5 are swept, that the quality factor Q_{L2} does not have a sharp maximum in the parameter space. Contrary, the maximum value $Q_{L2,\max}$ can be obtained relatively independent of the inductance L_2. This allows for first determining the maximum quality factor and afterwards for computing the inductance and capacitance which are demanded for maximum efficiency by equaling (3.38a) and (3.42), thus leading to

$$L_{2,\mathrm{opt}} = \frac{R_\mathrm{L}}{\omega_0 Q_{L2,\max}} \sqrt{1+k^2 Q_{L1} Q_{L2,\max}}, \qquad (3.43\mathrm{a})$$

$$C_{2,\mathrm{opt}} = \frac{1}{\omega_0^2 L_{2,\mathrm{opt}}}. \qquad (3.43\mathrm{b})$$

It should be mentioned that by using this definition, the optimum inductance L_2 is a function of the coupling factor as well as the properties of the primary coil due to the dependence of Q_{L1}.

Chapter 3 Inductive Power Transmission

In practical applications, the transponder must be operated in a variety of environments from which follows that the transponder antenna cannot be optimized for a certain coupling. A more practicable way is to design the inductance for zero coupling which gives a lower bound as

$$L_{2,\mathrm{min}} = \frac{R_\mathrm{L}}{\omega_0 Q_{L2,\mathrm{max}}}. \tag{3.44}$$

In this case, the optimum efficiency is obtained for low coupling. A further restriction is given in the case of mass-produced low-cost RFID transponder where the use of a discrete capacitor must be avoided, i.e. $C_{\mathrm{MN}} = 0$. In this case, the inductance is chosen in such a way to be resonant with

$$L_{2,\text{low-cost}} = \frac{1}{\omega_0^2 (C_\mathrm{p} + C_{\mathrm{Load}})}, \tag{3.45}$$

which is a direct consequence of the network topology according to Figure 3.12b. In addition, the quality factor Q_{L2} may be optimized by the geometrical parameters while keeping the inductance constant. Case studies of the explained design steps will be presented in section 6.2.3 on page 132. More details about the system design in case of RFID systems can be found in [70, 71].

Chapter 4
Partial Element Equivalent Circuit Method

In this chapter, the fundamental concepts of the Partial Element Equivalent Circuit (PEEC) method are derived. The PEEC method has been developed by Albert E. Ruehli in the early seventies [72, 73, 28] as a numerical approach for modeling the electromagnetic coupling effects of interconnecting structures. An important feature of the PEEC method is the fact that the electromagnetic field coupling effects of the structures are transferred to the circuit domain, represented as a system of lumped RLC network elements.[1] The transition to the network domain is obtained by partitioning the conductors of the analyzed interconnection structures into basic volume and surface cells with constant unknown currents and charges, respectively. The mutual EM interactions of the elements are interpreted in terms of partial resistances, inductances and capacitances which purely depend on the geometry and the material properties. These circuit elements are assembled to an equivalent circuit which can be solved via standard solvers such as SPICE or via linear algebra packages in both time and frequency domain. The network character of the system allows for an easy and straightforward integration of external circuit components.

In the last decades, the PEEC method has been extended by several authors and hence has become a general purpose numerical full wave and full spectrum[2] 3D method [75] with the possibility of including dielectric [76] and magnetic materials [77].

The chapter is organized as follows: First, the discretization of the fundamental equations is presented followed by the introduction of the partial network elements and the interpretation of the system as an equivalent circuit. Afterwards, the integration of different model simplifications such as quasi-stationary assumptions to the PEEC method is explained. The meshing of the interconnection structures is discussed in the subsequent section with special focus on IPT systems. The modeling of skin and proximity

---

[1] It should be mentioned at this point that similar considerations to interpret the MPIE formulation as a system of circuit elements has been presented by earlier authors such as Wessel in 1937 [74].

[2] The wording full spectrum refers to a numerical method which allows for an accurate modeling from DC up to a maximum frequency of interest which is only limited by the discretization.

effects via the PEEC method is discussed in a separate section due to the importance for the quality factors of PSCs. The chapter is closed with a brief overview of the modeling of dielectric and magnetic materials as well as acceleration techniques.

4.1 Discretization

As already mentioned in section 2.6, the system of equations (2.34) in the MPIE formulation is the basis for the PEEC method. In the following, all derivations are performed in the frequency domain whereas the transition to the time domain is straightforward, e. g. [78]. For simplicity reasons, the following analysis concentrates on an interconnection system located in free space. The inclusion of dielectric and/or magnetic materials will be focused on in section 4.7.

In free space, the total currents and charges are replaced by the currents and charges inside the conductors with $\vec{J}_{tot} = \vec{J}$ and $\varrho_{tot} = \varrho$ which allows for repeating (2.34) as

$$\frac{\vec{J}(\vec{r})}{\kappa(\vec{r})} + j\omega\mu_0 \int_{V'} \vec{J}(\vec{r}')\, G(\vec{r},\vec{r}')\, \mathrm{d}V' + \mathrm{grad}\,\Phi(\vec{r}) = 0, \tag{4.1a}$$

$$\frac{1}{\varepsilon_0} \int_{V'} \varrho(\vec{r}')\, G(\vec{r},\vec{r}')\, \mathrm{d}V' = \Phi(\vec{r}), \tag{4.1b}$$

$$\mathrm{div}\,\vec{J}(\vec{r}) + j\omega\varrho(\vec{r}) = 0. \tag{4.1c}$$

As before, (4.1a) is the MPIE formulation whereas the electric scalar potential Φ is defined by (4.1b). The continuity equation (4.1c) completes the EM system. Usually, the set of equations (4.1) is written as a system of two equations in which (4.1b) is substituted in (4.1a). The advantage of handling both equations separately is given by the fact that the MQS case is incorporated in the analysis since in this case, the system (4.1) is considered without (4.1b). Additionally, a better insight into the different basis and testing functions needed for the discretization of the above system is enabled.

According to section 2.6, the general Green's function $G(\vec{r},\vec{r}')$ has to be replaced by (2.32), depending on whether the full-wave solution or quasi-stationary assumptions are being used. The excitation of the system due to external sources may be introduced in form of an external electric field in the right hand side of (4.1a). Alternatively, virtual point current sources can be inserted in the right hand side of (4.1c). For simplicity reasons, these sources are not written explicitly in the following equations, instead they will be introduced in the network domain as voltage and current sources.

Although it has been shown in section 2.3.3 that for practical applications, the charges are located at the surfaces of the conductors only, a volume charge density is assumed in the following considerations. This is due to the fact that the intermediate steps of

4.1 Discretization

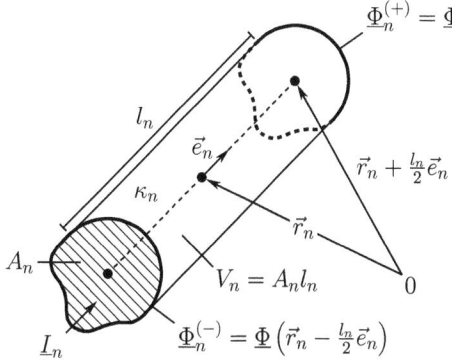

Figure 4.1: Basic PEEC current cell with the length l_n, the constant cross section A_n and conductivity κ_n as well as the homogeneous current density $\vec{J}_n = \underline{I}_n/A_n\,\vec{e}_n$. Both end faces are assumed as perfectly conducting which motivates to assign two nodes with constant potentials $\underline{\Phi}_n^{(-)}$ and $\underline{\Phi}_n^{(+)}$. These nodes provide interfaces to further charge and current cells.

the following derivations can be written in a more precise manner. In addition, the condition of zero charge density inside the conductors is already implied by the system (4.1), e. g. [78]. At the final stage of the formulation it will be explained how the volume charge density can be replaced by a surface charge density in order to reduce the complexity of the occurring integrals.

In order to recast (4.1) to a form which can be evaluated numerically, the conducting regions are discretized into a set of basic PEEC volume cells, accounting for the currents and charges which are the unknowns in the formulation. The overall number of current cells is denoted by $N_{\rm b}$ while the number of charge cells is denoted by $N_{\rm n}$, respectively. The subscripts "b" and "n" account for the branches and nodes in the network domain because each current cell will be represented by a branch in the circuit representation whereas each charge cell will be assigned to a single node.

A basic current cell is visualized in Figure 4.1. The volume of the cell V_n can be split into the arbitrary but constant cross section A_n and length l_n. The conductivity κ_n is required to be homogeneous in each cell and both end faces of the cell are assigned with a constant potential. A current $\underline{I}_n$ is flowing through the cell whereas the direction is pre-specified by the unit vector $\vec{e}_n$. This allows for modeling a conductor by a number of small current cells which are connected subsequently to their end faces.

Besides the presented basic current cell in Figure 4.1 usually being composed of a rectangular cross section, the PEEC method can also be derived for triangles (surface cells) and prisms as presented in [79, 80] for example. Moreover, the inclusion of volume cells with a varying cross section leads to non-orthogonal PEEC cells in general, which is discussed in [81, 82]. In cases where the direction of the current flow is not known a priori, e. g. for applications with ground planes, a 2D or 3D discretization of the structure as presented in section 4.5.1 is must be set up.

67

Chapter 4 Partial Element Equivalent Circuit Method

The volume cells for the charges are built equivalently to the current cells with the difference that the volume V_q of the cell is charged by the total charge $\underline{Q}_q$ and the cell is not directed.

In the standard PEEC method [28], the currents and charges are assumed to be constant inside each cell which allows for writing the total current and charge densities as a superposition of piecewise constant currents and charges according to

$$\vec{\underline{J}}(\vec{r}) = \sum_{n=1}^{N_\text{b}} \underline{I}_n \vec{m}_n(\vec{r}), \qquad \vec{m}_n(\vec{r}) = \begin{cases} \frac{\vec{e}_n}{A_n}, & \vec{r} \in \text{volume cell } n, \\ 0, & \text{otherwise}, \end{cases} \qquad (4.2\text{a})$$

$$\underline{\varrho}(\vec{r}) = \sum_{q=1}^{N_n} \underline{Q}_q v_q(\vec{r}), \qquad v_q(\vec{r}) = \begin{cases} \frac{1}{V_q}, & \vec{r} \in \text{volume cell } q, \\ 0, & \text{otherwise}. \end{cases} \qquad (4.2\text{b})$$

Besides the physically motivated description of the above approximation of the current and charge density, (4.2) can also be explained by the method of weighted residuals (cf. [22]). In this terminology, $\vec{m}_n$ and v_q are rectangular or piecewise constant orthogonal basis functions while $\underline{I}_n$ and $\underline{Q}_q$ are the unknown expansion coefficients. Since the method of weighted residuals is generally not restricted to this special kind of basis functions, some authors use different basis functions in the PEEC method, for instance, [22, 23, 78].

The motivation of the above approximation scheme (4.2) is to transfer the original unknown continuous current and charge distribution to a number of unknown coefficients which can be reformulated to a matrix system and be solved numerically via linear algebra methods. More precisely, substituting (4.2) into (4.1) and regarding the orthogonality of the basis functions $\vec{m}_n$ and v_q, which implies a reduction of the integrals to the volume of the n-th and q-th cell respectively, leads to the following form

$$\frac{1}{\kappa(\vec{r})} \sum_{n=1}^{N_\text{b}} \underline{I}_n \vec{m}_n(\vec{r}) + j\omega\mu_0 \sum_{n=1}^{N_\text{b}} \frac{\underline{I}_n \vec{e}_n}{A_n} \int_{V_n'} G(\vec{r},\vec{r}') \, dV' + \text{grad } \underline{\Phi}(\vec{r}) = 0, \qquad (4.3\text{a})$$

$$\frac{1}{\varepsilon_0} \sum_{q=1}^{N_n} \frac{\underline{Q}_q}{V_q} \int_{V_q'} G(\vec{r},\vec{r}') \, dV' = \underline{\Phi}(\vec{r}), \qquad (4.3\text{b})$$

$$\text{div} \sum_{n=1}^{N_\text{b}} \underline{I}_n \vec{m}_n(\vec{r}) + j\omega \sum_{q=1}^{N_n} \underline{Q}_q v_q(\vec{r}) = 0. \qquad (4.3\text{c})$$

In order to obtain a linear system of equations, the three equations above are tested via an inner product which is defined for two vector functions $\vec{a}$ and $\vec{b}$ as [26]

$$\langle \vec{a}(\vec{r}), \vec{b}(\vec{r}) \rangle = \int_V \vec{a}(\vec{r}) \cdot \vec{b}(\vec{r}) \, dV. \qquad (4.4)$$

4.1 Discretization

If the testing functions equal the basis functions, the methodology is called Galerkin method. An alternative is to use the collocation method which uses Dirac-delta distributions, e. g. [83].

Following the standard PEEC method, the Galerkin procedure is applied to the system (4.3) whereas the testing functions $\vec{m}_m$ and v_i are chosen equivalently to the basis functions. Thus,

$$\left\langle \vec{f}(\vec{r}),\ \vec{m}_m(\vec{r}) \right\rangle = \frac{1}{A_m} \int_{V_m} \vec{e}_m \cdot \vec{f}(\vec{r})\ \mathrm{d}V, \tag{4.5a}$$

$$\left\langle f(\vec{r}),\ v_i(\vec{r}) \right\rangle = \frac{1}{V_i} \int_{V_i} f(\vec{r})\ \mathrm{d}V, \tag{4.5b}$$

$$\left\langle f(\vec{r}),\ V_i v_i(\vec{r}) \right\rangle = \int_{V_i} f(\vec{r})\ \mathrm{d}V, \tag{4.5c}$$

with $\vec{f}(\vec{r})$ and $f(\vec{r})$ being the vector and scalar functions which have to be replaced by the equations (4.3). The above testing scheme with different normalizations is physically motivated by the aim to transform the EM field equations to the network domain where the unknowns are currents and voltages rather than current and charge densities. As an example, integrating the electric field terms of (4.3a) over the volume of the cell m and normalizing to the cross section of the cell as done in (4.5a), a typical voltage drop $U = \int \vec{E} \cdot \mathrm{d}\vec{s}$ over the cell is obtained.

When applying the above procedure (4.5) to the three equations in (4.3), the following set of equations is obtained

$$\frac{l_m}{\kappa_m A_m} I_m + j\omega\mu_0 \sum_{n=1}^{N_b} I_n \frac{\vec{e}_m \cdot \vec{e}_n}{A_m A_n} \int_{V_m}\int_{V_n'} G(\vec{r},\vec{r}')\,\mathrm{d}V'\,\mathrm{d}V +$$

$$+ \frac{1}{A_m} \int_{V_m} \vec{e}_m \cdot \mathrm{grad}\,\underline{\Phi}(\vec{r})\,\mathrm{d}V = 0, \tag{4.6a}$$

$$\frac{1}{\varepsilon_0} \sum_{q=1}^{N_n} \underline{Q}_q \frac{1}{V_i V_q} \int_{V_i}\int_{V_q'} G(\vec{r},\vec{r}')\,\mathrm{d}V'\,\mathrm{d}V = \frac{1}{V_i} \int_{V_i} \underline{\Phi}(\vec{r})\,\mathrm{d}V, \tag{4.6b}$$

$$\sum_{n=1}^{N_b} \underline{I}_n \int_{V_i} \mathrm{div}\,\vec{m}_n(\vec{r})\,\mathrm{d}V + j\omega \underline{Q}_i = 0. \tag{4.6c}$$

The last term of (4.6a) can be simplified by assuming that the potential of the m-th cell does not depend on the cross sectional dimensions, resulting in [28]

$$\frac{1}{A_m} \int_{V_m} \vec{e}_m \cdot \mathrm{grad}\,\underline{\Phi}(\vec{r})\,\mathrm{d}V = \frac{1}{A_m} \int_{A_m} \mathrm{d}A \int_{l_m} \mathrm{grad}\,\underline{\Phi}(\vec{r}) \cdot \mathrm{d}\vec{l}_m \tag{4.7a}$$

$$= \underline{\Phi}\left(\vec{r}_m + \tfrac{l_m}{2}\vec{e}_m\right) - \underline{\Phi}\left(\vec{r}_m - \tfrac{l_m}{2}\vec{e}_m\right) = \underline{\Phi}_m^{(+)} - \underline{\Phi}_m^{(-)} \tag{4.7b}$$

Chapter 4 Partial Element Equivalent Circuit Method

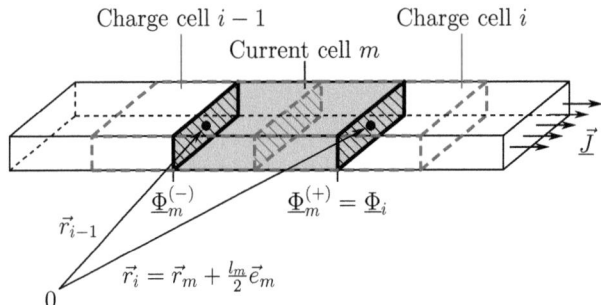

Figure 4.2: PEEC current and charge cells which are shifted by half of the cell length. This allows for matching the corresponding potentials located at the center points of the charge cells with the potentials at the end points of the current cells.

where $\underline{\Phi}_m^{(+)}$ and $\underline{\Phi}_m^{(-)}$ are the average potentials at the end faces of the cells as visualized in Figure 4.1.

According to the above considerations, the potential in the charge volume cell i is averaged by the right hand side term of (4.6b) and named $\underline{\Phi}_i$ in the following (cf. [73]). Substituting these simplifications into (4.6) and applying the Gauss' theorem to the volume integral of (4.6c), the equations can be reformulated as

$$\left[\frac{l_m}{\kappa_m A_m}\right]\underline{I}_m + j\omega \sum_{n=1}^{N_b}\left[\frac{\mu_0 \vec{e}_m \cdot \vec{e}_n}{A_m A_n}\int_{V_m}\int_{V_n'} G(\vec{r},\vec{r}')\,\mathrm{d}V'\,\mathrm{d}V\right]\underline{I}_n + \underline{\Phi}_m^{(+)} - \underline{\Phi}_m^{(-)} = 0, \quad (4.8\mathrm{a})$$

$$\sum_{q=1}^{N_n}\left[\frac{1}{\varepsilon_0 V_i V_q}\int_{V_i}\int_{V_q'} G(\vec{r},\vec{r}')\,\mathrm{d}V'\,\mathrm{d}V\right]\underline{Q}_q = \underline{\Phi}_i, \quad (4.8\mathrm{b})$$

$$\sum_{n=1}^{N_b}\underline{I}_n\int_{\partial V_i}\vec{m}_n(\vec{r})\cdot\mathrm{d}\vec{A} + j\omega\underline{Q}_i = 0. \quad (4.8\mathrm{c})$$

In order to match the potentials of the charge cells uniquely to the potentials at the end faces of the current cells, the cells are shifted by half of the length as visualized in Figure 4.2. Thus, each charge cell corresponds to a node which in turn belongs to an end point of the current cells. From this scheme it becomes obvious that the number of nodes N_n equals the number of charge cells.

The relative shift of the current and charge cells can be used to analyze the integral over the closed surface of (4.8c). Due to the dot product of the n-th current basis function with the outward normal vector of the charge-cell surface, only those parts of the surface have to be evaluated where the normal surface vector has either the same or opposite direction of the current flow in the neighboring current cell.

As can be seen in Figure 4.3, the value of the integral is only nonzero at the gray highlighted interfaces. Due to the carefully chosen normalization of the testing function

4.2 Partial Network Elements

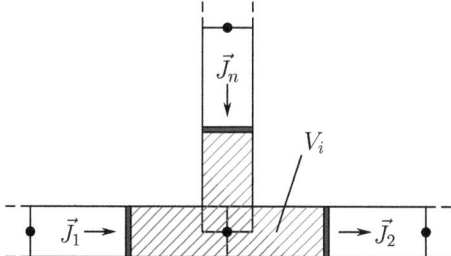

Figure 4.3: Connection of three current cells at a single node. The boundary integral (4.8c) over the current testing functions is nonzero only at the gray shaded interfaces.

(4.2a), the value of the integral is of the following form

$$B_{ni} = \begin{cases} +1, & \text{if current } n \text{ is leaving charge/node } i, \\ -1, & \text{if current } n \text{ is entering charge/node } i, \\ 0, & \text{remaining part.} \end{cases} \quad (4.9)$$

The entries B_{ni} belong to the sparse nodal connectivity or incidence matrix **B** with N_b rows and N_n columns. Note that just two nonzero entries exist in each row, one of them being +1 and the other −1 since each current cell is only connected to two nodes.

The above considerations allow to formulate the system (4.8) in a more compact form as

$$R_{mm}\underline{I}_m + j\omega \sum_{n=1}^{N_b} L_{mn}\underline{I}_n + \underline{\Phi}_m^{(+)} - \underline{\Phi}_m^{(-)} = 0, \quad (4.10a)$$

$$\sum_{q=1}^{N_n} P_{iq}\underline{Q}_q = \underline{\Phi}_i, \quad (4.10b)$$

$$\sum_{n=1}^{N_b} B_{ni}\underline{I}_n + j\omega\underline{Q}_i = 0, \quad (4.10c)$$

in which the bracketed terms of (4.8) have been abbreviated with R_{mm}, L_{mn} and P_{iq}. These terms will be interpreted as equivalent network elements with the exact definitions presented in the next section.

4.2 Partial Network Elements

The three expressions in the square brackets of (4.8) can be interpreted as partial network elements, i.e. partial coefficients of resistance, inductance and potential,[3] introduced by Ruehli [72, 73]. The coefficients are purely dependent on the geometry and

---
[3] The coefficient of potential is the reciprocal value of the capacitance according to (2.55).

Chapter 4 Partial Element Equivalent Circuit Method

the materials involved. When comparing the integrals with the basic definitions for resistance, inductance and coefficient of potential from section 2.7, the partial elements can be identified as being special cases of the general definitions.

In the following three paragraphs, the expressions are presented for the full-wave and quasi-stationary cases. Moreover, hints for the analytical evaluation of the corresponding integrals for special setups are given.

Partial Resistance The most trivial network coefficient is the partial resistance in the first square bracket of (4.8a)

$$R_{mm} = \frac{l_m}{\kappa_m A_m}, \qquad (4.11)$$

which corresponds to the standard expression for a conductor with the length l_m, the constant cross section A_m and conductivity κ_m according to the general definition (2.40). The subscripts mm indicate that the partial resistance contains only a self-term; different conductors are not coupled via mutual resistances. Due to the assumed homogeneous current density inside each current cell, the above equation can only account for the DC resistance of an interconnection structure. If an inhomogeneous current distribution should be captured, either the assumption of a homogeneous current density must be relaxed leading to different basis functions as in [23] or, alternatively, the cross section of the conductors must be subdivided which will be discussed in section 4.6.

Partial Inductance The second square bracket of (4.8a) can be interpreted as a partial mutual inductance between the volume cells m and n. When substituting the Green's function of (2.32), two different expressions are obtained

$$L_{mn} = \frac{\mu_0 \vec{e}_m \cdot \vec{e}_n}{4\pi A_m A_n} \int_{V_m} \int_{V_n'} \frac{1}{|\vec{r} - \vec{r}'|} dV' \, dV, \qquad \text{(LQS and MQS)} \qquad (4.12a)$$

$$\underline{L}_{mn}(j\omega) = \frac{\mu_0 \vec{e}_m \cdot \vec{e}_n}{4\pi A_m A_n} \int_{V_m} \int_{V_n'} \frac{e^{-jk|\vec{r} - \vec{r}'|}}{|\vec{r} - \vec{r}'|} dV' \, dV. \qquad \text{(Full wave)} \qquad (4.12b)$$

In contrast to the full-wave solution (4.12b), the partial inductance expression for the quasi-stationary approximations (4.12a) does not include frequency-dependent behavior. In this case, the partial inductance formulation coincides with the general expression (2.45) if three assumptions are fulfilled. First, the conductor's volume can be split into the length and the constant cross section. Second, the current density inside the conductors is homogeneous and third, the current flow does not change its direction inside the current cell.

The concept of partial inductances can be transferred to the commonly used inductance definition based on the magnetic flux through closed loops when the current cells

4.2 Partial Network Elements

are regarded as being a part of a virtually closed loop at infinity. In [72], this concept is discussed in detail.

It should be mentioned that two conductors which are oriented perpendicular to each other share zero mutual inductance. This is obvious due to the dot product $\vec{e}_m \cdot \vec{e}_n$ of both current cells in (4.12). Accordingly, if the two currents are flowing in the opposite direction, the mutual inductance becomes negative. A further property of the partial inductances is given by the fact that the coefficients are symmetrical. This can be ascribed to the Galerkin's procedure.

For practical applications, the six-fold integrals in (4.12) must be evaluated according to the specific geometry. Besides numerical integration techniques which are out of the scope of this work, exact analytical solutions exist for special arrangements, e. g. for parallel brick-shaped current cells. As an alternative, various approximation techniques can be applied. A detailed review about analytical techniques to solve (4.12a) is presented in appendix A.1 on page 157.

Partial Coefficient of Potential The partial coefficient of potential is defined as the bracketed expression of (4.8b) according to

$$P_{iq} = \frac{1}{4\pi\varepsilon_0 V_i V_q} \int_{V_i} \int_{V_q'} \frac{1}{|\vec{r} - \vec{r}'|} \, \mathrm{d}V' \, \mathrm{d}V, \qquad \text{(LQS)} \qquad (4.13a)$$

$$\underline{P}_{iq}(j\omega) = \frac{1}{4\pi\varepsilon_0 V_i V_q} \int_{V_i} \int_{V_q'} \frac{e^{-jk|\vec{r}-\vec{r}'|}}{|\vec{r} - \vec{r}'|} \, \mathrm{d}V' \, \mathrm{d}V. \qquad \text{(Full wave)} \qquad (4.13b)$$

As before, the expressions are symmetrical and only the quasi-stationary version is frequency independent. When comparing (4.13a) with the general definition for the coefficient of potential in (2.52), conformance is obtained in the case of a homogeneous charge density inside the volume.

In order to reduce the complexity of the above integrals, a common practice is to discretize the surfaces of the conductors only which is motivated by the vanishing charge density inside the conductors (s. section 2.3.3). This reduces the six-fold integral of (4.13) to a four-fold integral which is generally easier to compute. The transition from the volume to the surface involves a further subdivision of each volume charge cell into multiple panels being connected to a single node. In this case, the total number of panels is introduced as N_p.

As an example, the volume charge cell i as shown in Figure 4.2 can be replaced by four panels whereas the left and the right surfaces of the charge cell do not need to be considered as they do not carry charge. In practical applications with thin conductors, the thickness is usually neglected which allows for using a single surface only. As can be seen from section 6.2.3 on page 136, the obtained results are fairly accurate.

Chapter 4 Partial Element Equivalent Circuit Method

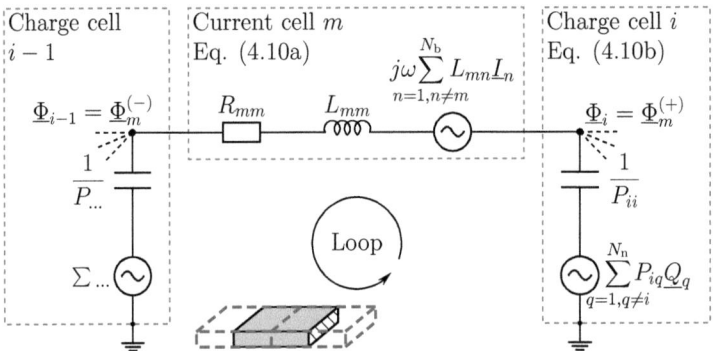

Figure 4.4: Equivalent RLC circuit representation of (4.10) with the m-th PEEC cell connected to the charge cells i and $i-1$.

In order to distinguish the surface from the cross section A in the mathematical notations, the surface of the i-th panel is indicated via S_i which allows for rewriting (4.13) for surface charges as

$$P_{iq} = \frac{1}{4\pi\varepsilon_0 S_i S_q} \int_{S_i}\int_{S_q'} \frac{1}{|\vec{r}-\vec{r}'|}\, \mathrm{d}A'\, \mathrm{d}A, \qquad \text{(LQS)} \qquad (4.14\mathrm{a})$$

$$\underline{P}_{iq}(j\omega) = \frac{1}{4\pi\varepsilon_0 S_i S_q} \int_{S_i}\int_{S_q'} \frac{e^{-jk|\vec{r}-\vec{r}'|}}{|\vec{r}-\vec{r}'|}\, \mathrm{d}A'\, \mathrm{d}A. \qquad \text{(Full wave)} \qquad (4.14\mathrm{b})$$

Again, the analytical solutions of (4.14a) for rectangular panels will be presented in appendix A.4 on page 164.

4.3 Equivalent Circuit Representation

In this section, the network character of the PEEC system will be emphasized by deducing Kirchhoff's Voltage Law (KVL) and Kirchhoff's Current Law (KCL) from the system of equations (4.10). Followed by that, the PEEC equations are converted to the matrix notation enabling to solve the system in the Modified Nodal Analysis (MNA) formulation.

The considerations start with pointing out the network character of the system (4.10). Figure 4.4 visualizes (4.10a) as a network branch consisting of the partial resistance R_{mm} and the partial self-inductance L_{mm}. A voltage source accounts for the mutual inductance interactions with other branch currents $\underline{I}_n$. The m-th branch of the network is

4.3 Equivalent Circuit Representation

connected to two nodes with the potentials being identical to the potentials of the connected charge cells (4.10b). Each charge cell can be represented by a partial capacitance in series with a further voltage source driven by all other charge cells q. The arrangement visualized in Figure 4.4 can be interpreted as KVL stating that the sum of all voltages around a closed loop is always zero. Accordingly, the third part (4.10c) describes the KCL which is a direct consequence of (4.9).

Due to the described equivalent circuit interpretation of the discretized EM system, it is possible to compute all partial network elements for a given application, store them together with the connectivity information (4.9) in a netlist, define ports and solve them with standard circuit solvers such as SPICE.

For the sake of completeness, the next section presents the basic steps required to obtain a linear system of equations which can be solved with standard linear algebra packages. The motivation is to obtain a deeper insight into the system and to provide the basis for the sensitivity analysis in chapter 5.

Generally, two different methods to solve electrical networks exist, the nodal and the mesh based analysis. The nodal method may be advantageous for systems with a small number of nodes compared to the number of branches [84]. For instance, this is the case for the single conductor example according to Figure 4.5, where multiple branches and panels are connected to a few nodes. The nodal based analysis has an additional implementation advantage because it is straightforward to set up the incidence matrix **B** of (4.9).

Contrary, when applying the mesh based analysis, a set of mesh currents is introduced which fulfills the KCL per definition because each mesh current enters and leaves a node at the same time [84]. Another advantage of the mesh based analysis is the fact that the obtained partial element matrices can directly be assembled to a symmetrical matrix and be solved for the mesh currents [85]. The drawback is the fact that a mesh matrix must be set up which requires additional algorithmic effort since the mesh matrix is not unique and different constraints must be maintained.

For this reason, the nodal based analysis is used throughout this work. The mesh based analysis is not discussed except for the extraction of the fast mutual inductance extraction technique of two multi-turn coils. In this specific case, the conductors are meshed as coarse as possible which makes the mesh based approach advantageous and involves a trivial mesh matrix as will be detailed in section 4.5.4.

4.3.1 Nodal Based Analysis

Until now, the PEEC system has only been written for a single cell in (4.10). The example of Figure 4.5 visualizes the discretization of a single conductor with multiple branches and panels connected to different nodes. The overall system description can be obtained by switching to the matrix formulation which captures all current and charge

Chapter 4 Partial Element Equivalent Circuit Method

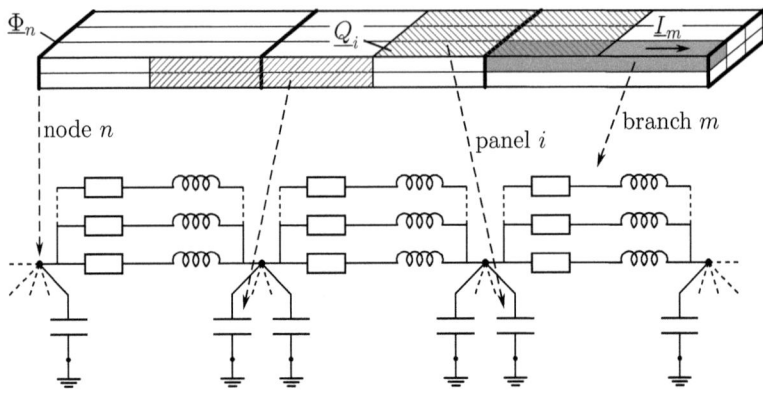

Figure 4.5: PEEC equivalent circuit of a single conductor with partial network elements. Multiple branches and panels are connected to each node. The mutual couplings of the network elements from Figure 4.4 are not visualized for clarity reasons.

cells at the same time. Although not visualized in Figure 4.5, generally each current cell is mutually coupled with all other current cells due to the sum term of (4.10a) whereas the same is true for the charges in (4.10b). Thus, (4.10) leads to a dense or full matrix system which can be expressed as[4]

$$(\mathbf{R} + j\omega\mathbf{L})\,\underline{\mathbf{i}}_{\mathrm{b}} - \mathbf{B}\,\underline{\boldsymbol{\varphi}}_{\mathrm{n}} = \underline{\mathbf{u}}_{\mathrm{s}}, \quad (4.15\mathrm{a})$$

$$\mathbf{P}_{\mathrm{p}}\,\underline{\mathbf{q}}_{\mathrm{p}} = \underline{\boldsymbol{\varphi}}_{\mathrm{p}}, \quad (4.15\mathrm{b})$$

$$\mathbf{B}^{\mathrm{T}}\underline{\mathbf{i}}_{\mathrm{b}} + j\omega\underline{\mathbf{q}}_{\mathrm{n}} = \underline{\mathbf{i}}_{\mathrm{s}}. \quad (4.15\mathrm{c})$$

In (4.15a), $\mathbf{R}$ is a $N_{\mathrm{b}} \times N_{\mathrm{b}}$ diagonal matrix consisting of partial resistances defined by (4.11). Equivalently, $\mathbf{L}$ is a $N_{\mathrm{b}} \times N_{\mathrm{b}}$ dense matrix composed of partial inductances according to (4.12). The vector $\underline{\mathbf{i}}_{\mathrm{b}}$ is the $N_{\mathrm{b}} \times 1$ branch current vector whereas the $N_{\mathrm{b}} \times 1$ voltage source vector $\underline{\mathbf{u}}_{\mathrm{s}}$ has been introduced in order to account for possible external voltage sources at the branches. The original difference of the potentials at the nodes $\underline{\Phi}_{m}^{(+)} - \underline{\Phi}_{m}^{(-)}$ from (4.10a) has been replaced by $-\mathbf{B}\,\underline{\boldsymbol{\varphi}}_{\mathrm{n}}$ which is a direct consequence of the introduced $N_{\mathrm{b}} \times N_{\mathrm{n}}$ incidence matrix $\mathbf{B}$ from (4.9). This is because each branch current enters and leaves exactly one node which leads to two nonzero entries in each row. Consequently, the $N_{\mathrm{n}} \times 1$ vector $\underline{\boldsymbol{\varphi}}_{\mathrm{n}}$ describes the node voltages.

When comparing the second equation (4.15b) with (4.10b), it can be verified that the subscripts have been changed to "p" in order to allow for multiple panels per node. As

---
[4]Throughout this work, matrices are written as bold uppercase letters whereas vectors are denoted by lowercase bold letters.

4.3 Equivalent Circuit Representation

before, $\mathbf{P}_\mathrm{p}$ is the $N_\mathrm{p} \times N_\mathrm{p}$ dense coefficient-of-potential matrix of the panels while $\mathbf{q}_\mathrm{p}$ as well as $\boldsymbol{\varphi}_\mathrm{p}$ are the $N_\mathrm{p} \times 1$ panel charge and potential vectors, respectively. In the last equation (4.15c), $\mathbf{q}_\mathrm{n}$ is the $N_\mathrm{n} \times 1$ vector of node charges. On the right hand side of the equation, the $N_\mathrm{n} \times 1$ current vector $\mathbf{i}_\mathrm{s}$ has been introduced in order to account for external current sources at the nodes. This is convenient when introducing ports to the network.

There exist a multitude of possibilities to solve the above matrix system, depending on how the equations are arranged, e.g. [78, 85, 86]. Here, the following procedure is chosen: In order to unify the different sized charge and potential vectors of (4.15) at the panels and the nodes, an $N_\mathrm{p} \times N_\mathrm{n}$ node reduction incidence matrix $\mathbf{D}$ is introduced similar to [85, 86] as

$$D_{mn} = \begin{cases} 1, & \text{if panel charge } m \text{ is connected to node } n, \\ 0, & \text{remaining part.} \end{cases} \qquad (4.16)$$

With the help of the node-reduction matrix, the following two relations can be enforced

$$\mathbf{q}_\mathrm{n} = \mathbf{D}^\mathrm{T} \mathbf{q}_\mathrm{p}, \qquad \boldsymbol{\varphi}_\mathrm{p} = \mathbf{D}\, \boldsymbol{\varphi}_\mathrm{n}, \qquad (4.17)$$

which state that all panel charges belonging to a single node are added while each node potential is copied to all adjacent panels in order to ensure a uniform potential on these panels. The relations are a consequence of the originally single charge volume cell partitioned into multiple surface cells. By means of (4.17), equation (4.15b) can be replaced by

$$\mathbf{q}_\mathrm{n} = (\mathbf{D}^\mathrm{T} \mathbf{P}_\mathrm{p}^{-1} \mathbf{D})\, \boldsymbol{\varphi}_\mathrm{n} = \mathbf{P}^{-1} \boldsymbol{\varphi}_\mathrm{n} = \mathbf{C}_\mathrm{s}\, \boldsymbol{\varphi}_\mathrm{n}. \qquad (4.18)$$

The $N_\mathrm{n} \times N_\mathrm{n}$ dense matrix $\mathbf{P}$ describes the node coefficients of potential. Its inverse $\mathbf{C}_\mathrm{s} = \mathbf{P}^{-1}$ is the $N_\mathrm{n} \times N_\mathrm{n}$ dense short-circuit capacitance matrix [76, 87]. In some applications, it might be advantageous to convert $\mathbf{C}_\mathrm{s}$ to the capacitance matrix $\mathbf{C}$ which is based on the voltages rather than the potentials. The conversion is straightforward and can be reviewed in [76, 87], for instance.

In order to set up the total matrix system, (4.15b) is replaced by (4.18). This equation, in turn, is substituted in (4.15c) which results together with (4.15a) in

$$\begin{bmatrix} \mathbf{R} + j\omega\, \mathbf{L} & \mathbf{B} \\ \mathbf{B}^\mathrm{T} & -j\omega\, \mathbf{C}_\mathrm{s} \end{bmatrix} \begin{bmatrix} \mathbf{i}_\mathrm{b} \\ -\boldsymbol{\varphi}_\mathrm{n} \end{bmatrix} = \begin{bmatrix} \mathbf{u}_\mathrm{s} \\ \mathbf{i}_\mathrm{s} \end{bmatrix}. \qquad (4.19)$$

The above system is written in the so-called Modified Nodal Analysis (MNA) formulation [88] because the unknowns are built by the branch currents together with the node potentials. The negative sign is included at the capacitance matrix as well as the nodal potentials in order to maintain a symmetric form of the system matrix assuming that the coefficient definitions from section 4.2 are used. The symmetrical property is advantageous for the adjoint sensitivity analysis specified in chapter 5 where the transposed system matrix is required. Due to the direct use of $\mathbf{C_s}$ in (4.19), the matrix inversion of $\mathbf{P_p}$ in (4.18) is required. This is accepted here since the size of the coefficient-of-potential matrix of IPT systems does usually not exceed a few hundreds of panels.[5] Whenever the direct inversion of $\mathbf{P_p}$ should be avoided, e. g. for large systems, alternative formulations such as presented in [78] can be applied.

4.3.2 Multi-Port Network

In this section, an expression for the port impedance matrix of a multi-port network is derived. The above system (4.19) is of the general form $\mathbf{A\,x} = \mathbf{b}$. Thus, it can be solved via standard linear algebra packages for the unknown branch currents and node potentials. In case of multi-port networks, it might be beneficial to define an expression for the port impedance matrix which can be used for the connection with external circuitry. To this end, an $N_\mathrm{n} \times N_\mathrm{port}$ sparse terminal incidence matrix $\mathbf{K}$ is introduced with N_port denoting the number of ports. The elements of the matrix are built by [78]

$$K_{mn} = \begin{cases} +1, & \text{if port current } n \text{ is entering node } m, \\ -1, & \text{if port current } n \text{ is leaving node } m, \\ 0, & \text{remaining part.} \end{cases} \quad (4.20)$$

This allows for reformulating the system (4.19) in the state-space form, such as presented in [41, 85]. Hence, a derivation of an impedance transfer function with the current source matrix $\underline{\mathbf{I}}_\mathrm{s} = \mathbf{K}\,\underline{I}_\mathrm{port}$ is obtained by

$$\begin{bmatrix} \mathbf{R} + j\omega\,\mathbf{L} & \mathbf{B} \\ \mathbf{B}^\mathrm{T} & -j\omega\,\mathbf{C_s} \end{bmatrix} \begin{bmatrix} \underline{\mathbf{I}}_\mathrm{b} \\ -\underline{\boldsymbol{\Phi}}_\mathrm{n} \end{bmatrix} = \begin{bmatrix} \mathbf{0} \\ \mathbf{K} \end{bmatrix} \underline{I}_\mathrm{port}, \quad (4.21\mathrm{a})$$

$$\underline{U}_\mathrm{port} = -\begin{bmatrix} \mathbf{0} & \mathbf{K}^\mathrm{T} \end{bmatrix} \begin{bmatrix} \underline{\mathbf{I}}_\mathrm{b} \\ -\underline{\boldsymbol{\Phi}}_\mathrm{n} \end{bmatrix}. \quad (4.21\mathrm{b})$$

Solving the first equation for the state variables and substituting the result into the second equation, the $N_\mathrm{port} \times N_\mathrm{port}$ port impedance matrix is obtained as

$$\underline{\mathbf{Z}}_\mathrm{port} = \frac{\underline{U}_\mathrm{port}}{\underline{I}_\mathrm{port}} = -\begin{bmatrix} \mathbf{0} & \mathbf{K}^\mathrm{T} \end{bmatrix} \begin{bmatrix} \mathbf{R} + j\omega\,\mathbf{L} & \mathbf{B} \\ \mathbf{B}^\mathrm{T} & -j\omega\,\mathbf{C_s} \end{bmatrix}^{-1} \begin{bmatrix} \mathbf{0} \\ \mathbf{K} \end{bmatrix}. \quad (4.22)$$

4.4 Model Simplifications

This section concentrates on adapting the general derivations of the PEEC method from the previous sections to different approximation techniques of the Maxwell's equations

---
[5] A relatively coarse capacitive mesh is sufficient in IPT systems because capacitive cross coupling is dominated by the inductive effects.

such as the quasi-stationary assumptions. In the first subsection, some hints about the full-wave and LQS-PEEC systems will be given. Afterwards, two simplifications will be discussed, the MQS-PEEC approach as well as the DC-PEEC limit which corresponds to the formulation of stationary currents. The section is concluded by a short review of the 2D-PEEC formulation which uses an adapted Green's function.

4.4.1 Full-Wave (rPEEC) and Quasi-Stationary (LQS-PEEC)

The preceding derivation of the PEEC method is valid for the full-wave as well as the LQS case assuming that the interconnection structure is surrounded by a homogeneous medium. In the following, the two models are named Retarded Partial Element Equivalent Circuit (rPEEC) according to [75] and Lorenz-Quasi-Static – Partial Element Equivalent Circuit (LQS-PEEC) motivated by section 2.4.1. The latter model is also referred to as (L_p, P, R)-PEEC in some contributions, e.g. [41].

Both models do only differ in the expression for the partial inductances (4.12) and partial coefficients of potential (4.13) or (4.14). In contrast to the integrals in the LQS formulation, the full-wave elements are frequency dependent and generally complex valued. The reason is the exponential function in the integral kernels causing a phase term of the mutual element couplings due to the finite speed of light. In time-domain formulations, this corresponds to a retardation term. Since the retardation term generally complicates the integral evaluations, the full-wave integrals are avoided whenever possible, i.e. when analyzing applications with dimensions much smaller than the minimum wavelength. Since the geometrical dimensions and the frequency spectrum of IPT systems typically fulfill these requirements, quasi-stationary assumptions are justified. For this reason, only a few hints about the full-wave integrals are presented in the following.

For applications with dimensions comparable to the wavelength, the geometry is typically discretized with 10 to 20 cells per wavelength, e.g. [42, 89]. As stated in [75], no closed-form solutions of the retarded coefficient integrals exist. This demands for a numerical evaluation of the integrals for each frequency point in general. In order to overcome this drawback, some authors move the exponential phase term out of the integral [89], thus storing a frequency-dependent phase term between the center points of two elements [90]. As an alternative, in [83], the testing procedure is performed with Dirac-delta shaped testing distributions in order to reduce the complexity of the partial element integrals.

4.4.2 Magneto-Quasi-Static (MQS-PEEC)

In this section, the Magneto-Quasi-Static – Partial Element Equivalent Circuit (MQS-PEEC) method, which is also referred to as (L_p, R)-PEEC approach, is focused on. As already mentioned in section 2.6, the MPIE formulation differs for MQS and LQS systems in a basic property. In the MQS case, the continuity equation has to be replaced by div $\vec{\underline{J}} = 0$ from (2.29) which results in a decoupling of the currents from the charges. This allows the omission of the charges. More precisely, the fundamental MPIE system of equations (4.1) simplifies in the MQS case to

$$\frac{\vec{\underline{J}}(\vec{r})}{\kappa(\vec{r})} + j\omega\mu_0 \int_{V'} \vec{\underline{J}}(\vec{r}')\, G(\vec{r},\vec{r}')\, \mathrm{d}V' + \operatorname{grad}\underline{\Phi}(\vec{r}) = 0, \qquad (4.23\mathrm{a})$$

$$\operatorname{div}\vec{\underline{J}}(\vec{r}) = 0. \qquad (4.23\mathrm{b})$$

Instead of repeating the entire derivation for this modified system of equations, only the differences w.r.t. the preceding derivation are emphasized. When discretizing the above system with the same basis and testing functions as in section 4.1, (4.10) results in

$$R_{mm}\underline{I}_m + j\omega \sum_{n=1}^{N_b} L_{mn}\underline{I}_n + \underline{\Phi}_m^{(+)} - \underline{\Phi}_m^{(-)} = 0, \qquad (4.24\mathrm{a})$$

$$\sum_{n=1}^{N_b} B_{ni}\underline{I}_n = 0. \qquad (4.24\mathrm{b})$$

The above equations again describe the KVL and KCL with the same nodal connectivity matrix of (4.9) as well as the partial resistances and inductances according to (4.11) and (4.12a). The coefficients of potentials do not have to be considered since the charges do not influence the overall system behavior. When expressing (4.24) in the MNA matrix notation and repeating the steps of section 4.3, the following linear system of equations is obtained (cf. also [30, eq. (16)])

$$\begin{bmatrix} \mathbf{R} + j\omega\,\mathbf{L} & \mathbf{B} \\ \mathbf{B}^{\mathrm{T}} & \mathbf{0} \end{bmatrix} \begin{bmatrix} \mathbf{i}_\mathrm{b} \\ -\underline{\boldsymbol{\varphi}}_\mathrm{n} \end{bmatrix} = \begin{bmatrix} \underline{\mathbf{u}}_\mathrm{s} \\ \underline{\mathbf{i}}_\mathrm{s} \end{bmatrix}. \qquad (4.25)$$

If this MQS system is compared with the LQS version of (4.19), the only difference is observed in leaving out the capacitance matrix $\mathbf{C}_\mathrm{s}$.[6] This fact motivates for a joint simulation of the LQS-PEEC and MQS-PEEC models since the partial element integrals have to be computed just once. In other words, the only overhead is to solve two systems which often constitutes a minor part of the overall simulation time only. Further properties of the joint simulation are the same element discretization and solution accuracy. This makes the technique ideally suited for the network model extraction of PSCs as already explained in section 3.3.2.

---

[6] In the MQS case, one of the network nodes must be defined as the reference node [84] to which the node potentials $\underline{\boldsymbol{\varphi}}_\mathrm{n}$ can be referred to.

4.4 Model Simplifications

4.4.3 Stationary Currents (DC-PEEC)

The Direct Current – Partial Element Equivalent Circuit (DC-PEEC) or simply R-PEEC method can be regarded as the DC limit of the MQS-PEEC approach from above. In this case, the system (4.25) reduces to the sparse tableau form

$$\begin{bmatrix} \mathbf{R} & \mathbf{B} \\ \mathbf{B}^{\mathrm{T}} & \mathbf{0} \end{bmatrix} \begin{bmatrix} \mathbf{i}_{\mathrm{b}} \\ -\boldsymbol{\varphi}_{\mathrm{n}} \end{bmatrix} = \begin{bmatrix} \mathbf{u}_{\mathrm{s}} \\ \mathbf{i}_{\mathrm{s}} \end{bmatrix}. \quad (4.26)$$

The DC-PEEC model is used in section 6.2.1 for analyzing the DC resistance of a rectangular conductor bend for which an analytical reference solution is available.

4.4.4 2D Magneto-Quasi-Static (2D-PEEC)

In this section, the Two Dimensional – Partial Element Equivalent Circuit (2D-PEEC) method is introduced as being a 2D version of the MQS-PEEC approach. Results of a 2D-PEEC analysis will be evaluated in section 6.1.2, where eddy-current problems of a cylindrical conductor are analyzed and compared with analytical expressions. The derivation of the 2D-PEEC method is motivated in [91] and can be shown to be a 2D counterpart of the MQS-PEEC method. The basic difference is the fact that the Green's function $\hat{G}(\vec{r}, \vec{r}')$ of (2.32) has to be replaced by its 2D counterpart

$$\hat{G}_{\mathrm{2D}}(\vec{r}, \vec{r}') = -\frac{1}{2\pi} \ln\left(|\vec{r} - \vec{r}'|\right). \quad (4.27)$$

Following the derivations of section 4.1 but modifying the testing procedure by surface integrals, the partial per-unit-length resistance is obtained according to (4.11) as

$$R'_{mm} = \frac{1}{\kappa_m A_m}. \quad (4.28)$$

Similarly, the partial per-unit-length inductances from (4.12a) in the 2D case are obtained as [92]

$$L'_{mn} = -\frac{\mu_0}{2\pi A_m A_n} \int_{A_m} \int_{A'_n} \ln\left(\sqrt{(x-x')^2 + (y-y')^2}\right) \mathrm{d}x' \, \mathrm{d}y' \, \mathrm{d}x \, \mathrm{d}y. \quad (4.29)$$

Analytical solutions for the integrals of (4.29) for elements with rectangular cross section will be presented in appendix A.3 on page 164.

According to the MQS-PEEC case, the linear system of equations (4.25) remains unchanged. Special attention must be paid to 2D applications where the total sum of all currents is nonzero. This is due to the logarithmic character of the Green's functions [92] which implies an infinite total magnetic energy as well as an infinite total inductance [93], respectively. For this reason, either the internal inductance or inductance differences are commonly evaluated in such cases [91, 94].

4.5 Meshing Strategies

Different meshing strategies of PEEC systems are detailed in this section whereas it is focused on those mesh settings which are needed for the modeling of PSCs. The section is structured as follows: After briefly discussing the complexity of different mesh types, the discretization of rectangular conductor bends is focused on by applying three different meshing techniques. Afterwards, the panel mesh of a rectangular PSC with thin conductors is discussed. The section is closed by the presentation of an efficient technique to extract the mutual inductance of two arbitrarily shaped and positioned spiral coils by using the most simple mesh setting. It will be shown that this technique corresponds to the Greenhouse method [53].

Although the subject of meshing the cross sections of the conductors belongs to this section, it is shifted to section 4.6 since it constitutes a crucial point in IPT system design.

4.5.1 1D, 2D and 3D Meshes

For arbitrary 3D conducting structures where the current direction is not known a priori, the geometry needs to be meshed with a 3D grid of nodes and a volume current cell in between each pair of neighboring nodes. In addition, the surface of the structure must be meshed with a set of panels, at least a single panel per surface node. Due to the fact that each non-orthogonal current pair is mutually coupled via (4.12) and every charge pair via (4.14), the system matrix in (4.19) is densely populated. Thus, a direct solution grows as $O(N^3)$ in time while the matrix storage grows as $O(N^2)$, respectively, with N representing the number of unknowns [42]. From this fact it can be concluded that the PEEC method has its main advantage for interconnection structures with long and thin wires, where the unknown currents can be limited to the direction of the estimated current flow. Hereby, the number of unknowns is remarkably reduced.

An exemplary mesh of a PSC is visualized in Figure 4.6 as a mixture of 1D and 2D regions. An adequate meshing algorithm first extracts parts of the structure which can be cast into straight segments leading to regions with a 1D discretization. The remaining part is discretized with a set of nodes whereas two neighboring nodes are connected via a current cell each. In Figure 4.6, the cells are composed of rectangular bricks. A more precise modeling for arbitrary curved objects can be obtained by using non-orthogonal elements as presented in [81, 82].

In contrast to the 2D partitioning, the 1D discretization of the conductors is performed with the lowest possible number of straight segments leading to long and thin volume cells.[7] The eventually occurring high aspect ratios of the lengths and cross sectional dimensions are not problematic as long as the analytical integral representations

---

[7] If the lengths of the individual cells are longer than 1/20 to 1/10 of the wavelength, a further subdivision is required.

4.5 Meshing Strategies

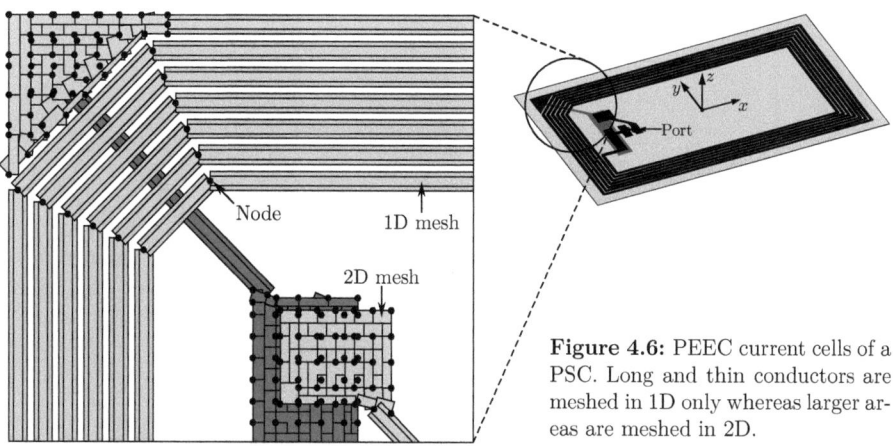

Figure 4.6: PEEC current cells of a PSC. Long and thin conductors are meshed in 1D only whereas larger areas are meshed in 2D.

specified in appendix A.1 are being used. The cross sections of the segments can further be subdivided in order to account for frequency-dependent eddy-current losses which will be focused on in section 4.6. Prior to that, the discretization of the corners of the conductors is discussed.

4.5.2 Discretization of Conductor Bends

Although the current direction follows the direction of the conductors in general, special attention must be paid to the conductor bends where the conductors change the direction. Generally, the change of current flow is of curved, smooth nature. Despite of this fact, for maintaining straight segments and for keeping the numerical effort as low as possible, typically only a single node is put at each corner to which two or more straight current cells are connected to. In order to quantify the error of such an approach, different discretization routines of corners are compared. Special attention is paid towards rectangular corners since they often occur in rectangular PSCs.

The discretization of rectangular conductor bends can be performed according to the sketched variants shown in Figure 4.7. Because the current density is extremely high at the innermost edge (s. Figure 6.15 on page 125), a proper discretization scheme to capture the 2D current distribution as visualized in Figure 4.7a is mandatory to achieve a high accuracy. Contrary, a simplified discretization scheme as shown in Figure 4.7b can be set up, where the sum of all currents is forced to be zero at a single node. A third discretization scheme is presented in Figure 4.7c, where the overlapping areas of the second variant are avoided by adapting the length of each segment according to the relative position. The number of elements is identical to the variant in Figure 4.7b since only one node is introduced at each corner. The number of unknown currents is of order

Chapter 4 Partial Element Equivalent Circuit Method

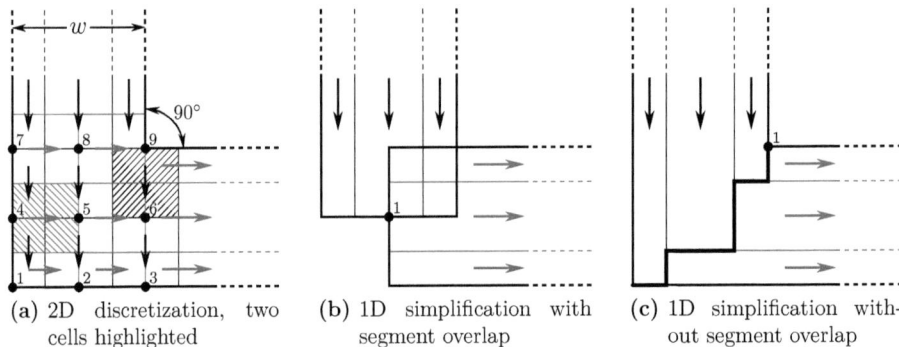

(a) 2D discretization, two cells highlighted

(b) 1D simplification with segment overlap

(c) 1D simplification without segment overlap

Figure 4.7: Different discretization schemes to represent the 90° corners of the conductors. The arrows indicate the direction of current flow in each cell. Only the discretization method (a) is able to reflect the non-abrupt change of current direction correctly whereas the gain of accuracy is compensated by the numerical effort. In the versions (b) and (c), the corner is equipped with a single node each whereas an unphysical overlapping of the elements is avoided in version (c).

$O(N_w^2)$ for the discretized corner and of order $O(N_w)$ for the simplified versions with N_w being the number of subdivisions in the conductor width. Numerical results for all three approaches will be presented in section 6.2.2.

In the following, the 2D meshing algorithm according to Figure 4.7a will be discussed in more detail as it is intensively used for a convergence study in section 6.2.1. An exemplary discretization of a rectangular conductor bend is presented in Figure 4.8 in which the elements are only visualized with 50 % of their actual width in order to allow a distinction between x- and y-directed cells. The conductors themselves are subdivided with seven non-equidistant bars in order to account for the estimated non-uniform current distribution (s. section 4.6.2). Starting with the 2D discretization at a distance Δl_i towards the unconnected conductor ends (cf. Figure 4.8), the change of current direction can be accounted for in an accurate manner. It should be mentioned that the current cells at the conductor edges have only half of the width compared to the cells in the interior of the conductors which is discussed detailed in [95].

4.5.3 Panel Mesh of a Printed Spiral Coil with Thin Conductors

The relatively coarse but efficient panel discretization of a rectangular PSC is focused on in this section. The use of such a mesh is motivated by the fact that in PSCs, the electric energy plays only a minor role in the overall system behavior. This motivates to neglect the thickness of the conductors and to treat the structure in a 2D manner

4.5 Meshing Strategies

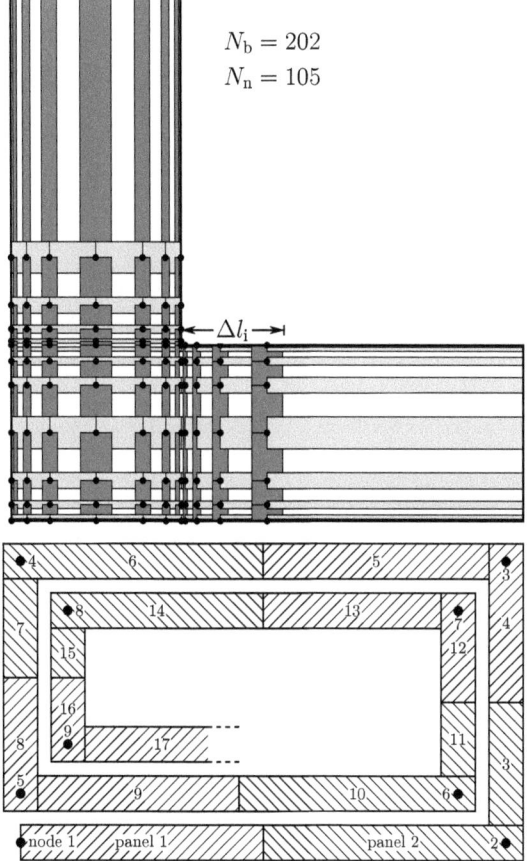

$N_\mathrm{b} = 202$
$N_\mathrm{n} = 105$

Δl_i

Figure 4.8: PEEC discretization of a rectangular conductor bend with 202 branches and 105 nodes. The current cells are only visualized with 50 % of their actual width in order to allow a distinction between x- and y-directed cells. The 2D discretization is enlarged by the length Δl_i towards the direction of the long conductors in order to more precisely account for the change of direction of the estimated current flow.

Figure 4.9: Capacitive 2D cells of a rectangular PSC. Each corner of the coil is equipped with a single node whereas the straight conductors are halved for the panel discretization. Each node is assigned to the two panels which are located most closely.

which is justified if the width of the conductors is much larger than the thickness. This leads to a significant reduction of the number of unknowns because a surface mesh of the side panels as well as the bottom panel is avoided.

The capacitive discretization scheme of the conducting sheets is presented in Figure 4.9, visualized with one cell per conductor width.[8] A single node is attached at each corner of the coil according to the corner discretization variants of Figures 4.7b and 4.7c. The lengths of the conductors are divided into halves in order to ensure rectangular patches and to allocate the potentials to the circuit nodes at the corners which

---

[8]In order to capture a possibly non-uniform charge distribution along the conductor's widths, the surface can be further discretized according to the scheme as presented in Figure 4.11.

85

Chapter 4 Partial Element Equivalent Circuit Method

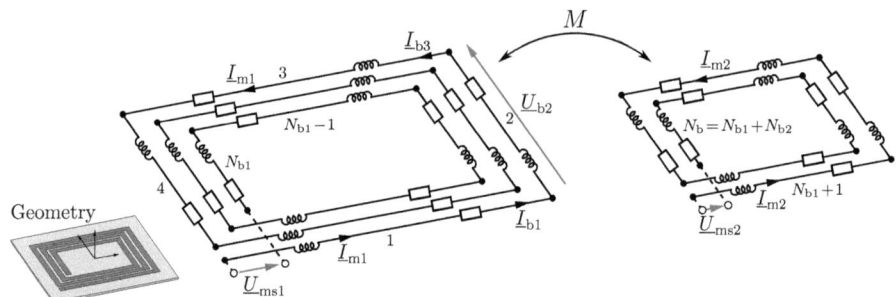

Figure 4.10: Mutual inductance computation of two inductively coupled PSCs. The coils are modeled by straight filaments. Each filament is represented by two lumped RL elements whereas the mutual couplings of the elements are not visualized in this figure. Two mesh currents have been introduced which flow through each of the coils. In this most simple topology with a single branch per straight conductor, the mutual inductance of both coils is obtained by summing the partial mutual inductances according to (4.35) instead of solving a system of equations.

is demanded for the node reduction matrix of (4.16). This typically leads to less than a hundred panels for a multi-turn coil.

4.5.4 Mutual Inductance Computation of two Spiral Coils

This section proposes a technique for the fast and efficient computation of the mutual inductance between two PSCs which is needed for the transformer-concept modeling described in section 3.3.1. In contrast to the self-impedance extraction of each coil, where an accurate internal conductor modeling is required for a precise loss prediction (s. section 4.6), the mutual inductance computation should preferably be performed by a fast extraction technique. This is even more motivated by the aim to allow for fast spatial sweeps, e. g. to enable a forecast of the maximum powering range.

To enable a fast simulation speed, the inductively coupled coils are discretized with a filamentary set of straight conductors according to Figure 4.10 whereas each straight segment between two nodes is modeled by a single filament only. The first inductor is represented by N_{b1} branches (current cells) and the second by N_{b2} branches, respectively, leading to the number of branches $N_b = N_{b1} + N_{b2}$ in total. The MQS-PEEC method[9] is applied to the structure. Afterwards, the $N_b \times N_b$ matrices of partial resistances[10] **R** and inductances **L** are set up.

---

[9]The MQS-PEEC method is a reasonable choice since the inductance concept of closed current loops is based on the MQS assumption.

[10]The partial resistances are not necessarily required as will be seen in (4.35) but are included here for reasons of completeness.

4.5 Meshing Strategies

In the following derivation, the network system will be analyzed in the mesh based approach, in which a set of mesh currents is introduced whereas each mesh current forms a closed current loop in the network topology, e.g. [84]. The mesh currents identically satisfy the KCL since each mesh current enters and leaves a node at the same time. The mesh based approach is advantageous if the number of nodes and branches are comparable [30] which is the case in this example, because only two mesh currents are required. For applying the mesh based analysis to the MQS-PEEC system, (4.15a) is rewritten in a modified manner as

$$(\mathbf{R} + j\omega \mathbf{L})\,\underline{\mathbf{i}}_b = \underline{\mathbf{u}}_b, \tag{4.30}$$

whereas the external branch sources $\underline{\mathbf{u}}_s$ of (4.15a) are not regarded in the formulation. The $N_b \times 1$ vector of branch voltages $\underline{\mathbf{u}}_b = \mathbf{B}\,\underline{\boldsymbol{\varphi}}_n$ has been introduced in (4.30) in order to avoid to set up the nodal connectivity matrix. By sorting the branches of each coil as depicted in Figure 4.10, the vector of branch currents $\underline{\mathbf{i}}_b = [\underline{\mathbf{i}}_{b1}, \underline{\mathbf{i}}_{b2}]^T$ and the vector of branch voltages $\underline{\mathbf{u}}_b = [\underline{\mathbf{u}}_{b1}, \underline{\mathbf{u}}_{b2}]^T$ can be separated into the parts belonging to the individual coils.

In order to extract the port impedance matrix of both coils it is necessary to set up a mesh matrix $\mathbf{M}$. While the setup of the mesh matrix is generally not unique [84], in the specific case specified in Figure 4.10 it is composed of two rows only

$$\mathbf{M} = \begin{bmatrix} M_{11} & \cdots & M_{1N_{b1}} & M_{1N_{b1}+1} & \cdots & M_{1N_b} \\ M_{21} & \cdots & M_{2N_{b1}} & M_{2N_{b1}+1} & \cdots & M_{2N_b} \end{bmatrix} = \begin{bmatrix} 1 & \cdots & 1 & 0 & \cdots & 0 \\ 0 & \cdots & 0 & 1 & \cdots & 1 \end{bmatrix}. \tag{4.31}$$

The matrix $\mathbf{M}$ relates the terminal voltages to the branch voltages and superposes all mesh currents flowing through a specific branch to the corresponding branch current according to [30]

$$\mathbf{M}\,\underline{\mathbf{u}}_b = \underline{\mathbf{u}}_{ms}, \qquad \mathbf{M}^T \underline{\mathbf{i}}_m = \underline{\mathbf{i}}_b, \tag{4.32}$$

where $\underline{\mathbf{u}}_{ms} = [\underline{U}_{ms1}, \underline{U}_{ms2}]^T$ is the mesh source voltage vector and $\underline{\mathbf{i}}_m = [\underline{I}_{m1}, \underline{I}_{m2}]^T$ the mesh current vector. Combining (4.30) and (4.32) results in

$$\left[\mathbf{M}\,(\mathbf{R} + j\omega \mathbf{L})\,\mathbf{M}^T\right] \underline{\mathbf{i}}_m = \underline{\mathbf{u}}_{ms}. \tag{4.33}$$

This equation relates the port voltages with the mesh currents which are identical to the port currents of the two coupled coils. As can be verified by Figure 4.10, the expression in the square brackets of (4.33) is the 2×2 port impedance matrix according to the general definition from (4.22). When separating this matrix into real and imaginary parts, the real-part matrix can be shown to have only two nonzero entries on the main diagonal. These elements are the self-resistances of both coils. Contrary, the evaluation of the imaginary-parts results in a 2×2 symmetrical inductance matrix. Consequently,

Chapter 4 Partial Element Equivalent Circuit Method

the mutual inductance M of both conductors is represented by the 12-element or 21-element of the matrix $\mathbf{M L M^T}$ according to

$$\mathbf{M L M^T} = \begin{bmatrix} 1 & 0 \\ \vdots & \vdots \\ 1 & 0 \\ 0 & 1 \\ \vdots & \vdots \\ 0 & 1 \end{bmatrix}^T \begin{bmatrix} L_{1,1} & \cdots & L_{1,N_{b1}} & L_{1,N_{b1}+1} & \cdots & L_{1,N_b} \\ \vdots & \ddots & \vdots & \vdots & \ddots & \vdots \\ L_{N_{b1},1} & \cdots & L_{N_{b1},N_{b1}} & L_{N_{b1},N_{b1}+1} & \cdots & L_{N_{b1},N_b} \\ L_{N_{b1}+1,1} & \cdots & L_{N_{b1}+1,N_{b1}} & L_{N_{b1}+1,N_{b1}+1} & \cdots & L_{N_{b1}+1,N_b} \\ \vdots & \ddots & \vdots & \vdots & \ddots & \vdots \\ L_{N_b,1} & \cdots & L_{N_b,N_{b1}} & L_{N_b,N_{b1}+1} & \cdots & L_{N_b,N_b} \end{bmatrix} \begin{bmatrix} 1 & 0 \\ \vdots & \vdots \\ 1 & 0 \\ 0 & 1 \\ \vdots & \vdots \\ 0 & 1 \end{bmatrix} \quad (4.34a)$$

$$= \begin{bmatrix} \sum_{m=1}^{N_{b1}} \sum_{n=1}^{N_{b1}} L_{mn} & \sum_{m=1}^{N_{b1}} \sum_{n=N_{b1}+1}^{N_{b1}+N_{b2}} L_{mn} \\ \sum_{m=N_{b1}+1}^{N_{b1}+N_{b2}} \sum_{n=1}^{N_{b1}} L_{mn} & \sum_{m=N_{b1}+1}^{N_{b1}+N_{b2}} \sum_{n=N_{b1}+1}^{N_{b1}+N_{b2}} L_{mn} \end{bmatrix}. \quad (4.34b)$$

When comparing the result of (4.34b) with (4.33), the expected 2×2 port inductance matrix is obtained. Thus, the mutual inductance is expressed by element 12 of the matrix in (4.34b) as [70]

$$M = \sum_{m=1}^{N_{b1}} \sum_{n=N_{b1}+1}^{N_{b1}+N_{b2}} L_{mn}. \quad (4.35)$$

In contrast to the matrix system (4.22) which has to be solved for the port impedances, the simple approach with a single branch in between two nodes requires only the evaluation of $N_{b1} N_{b2}$ partial inductances as well as their summation according to (4.35), thus drastically reducing the computational effort. For visualization aspects, the necessary partial mutual inductances have been gray shaded in (4.34a). The simulation effort can even more be reduced by approximating the volume integrals of the partial inductances of (4.12a) by line integrals when filamentary currents are assumed. This allows for using approximated integral expressions, justified by two reasons: First, two segments from different coils are usually located at a reasonable distance in which the differences between the volume and line integrals vanish. Second, no full solutions of the integrals are required as the concept of closed current loops does not take into account the internal inductance. Another advantage of the filamentary approach is the fact that equations exist for an arbitrary orientation of the segments (cf. [54]).

The above presented results are equivalent to the Greenhouse method [53] in which generally each conductor is discretized by a single segment, hence avoiding to solve a linear system of equations. In the Greenhouse method, also the self-inductance expressions on the main diagonal of (4.34b) are utilized. Since the simple mesh does not allow for the inclusion of internal conductor effects such as frequency-dependent losses, these main diagonal terms are not used in this work. However, the self-inductance expressions of (4.34b) may be used as a starting point for design and optimization purposes.

4.6 Modeling of Skin and Proximity Effects

Due to the importance of the loss modeling in IPT systems, a dedicated section has been introduced accounting for this subject. First, a brief review about the state-of-the-art techniques of the modeling of skin- and proximity-effect losses is presented. It is motivated that the classic volume discretization is appropriate for IPT systems. Followed by this, the subdivision technique of the conductor's cross section is detailed for the rectangular and the circular cross sections.

4.6.1 State-of-the-Art Techniques

In order to account for an inhomogeneous current distribution inside conductors occurring at high frequencies due to induced eddy currents, the meshing of each conductor with a single PEEC cell according to Figure 4.1 does not adequately represent the physical behavior. Instead, the DC resistance is obtained independent of the chosen frequency. In order to overcome this limitation, a commonly used approach [30, 72] is to model each conductor by a bundle of parallel-connected rectangular basic cells which are also referred to as bricks. The mutual interactions of the bricks are accounted for by partial inductances as before. As a result, the inhomogeneous current distribution is approximated in a stair-case manner. Exemplary results are shown in Figure 6.4 for a circular cross section.

The described standard technique allows the consideration of any frequency-dependent current distribution since the eddy currents themselves are modeled due to the mutual inductance interactions. The main drawback of this approach is the fact that the element size must be chosen comparable to the skin depth at the highest frequency of interest in order to obtain a sufficient accuracy. Thus, the method is inefficient at high frequencies where the current is mainly concentrated at the surface of the conductors and an adequate meshing becomes cumbersome. A common way to reduce the complexity in such cases is to mesh the cross section with non-equidistant segments resulting in a lower number of elements, especially at the interior of the conductor where the gradient of the current density is low.

In order to completely avoid the intensive volume meshing at high frequencies, the common approach is to approximate the volume currents by equivalent surface currents. Hereby, the need for resolving the interior is eliminated. In this case, the losses are accounted for by introducing a frequency-dependent surface impedance which afterwards replaces the standard DC resistance in the PEEC cells displayed in Figure 4.4. The surface impedances can either be computed numerically in a per-unit-length manner as presented in [96] or, alternatively, by introducing a subnetwork [22, 66] which approximates the skin influence by a ladder-type network of lumped elements. In the latter approach, a parameter fitting technique is applied which is similar to the coil impedance macromodeling according to section 3.3.2.2, thus allowing for simulations in both time and frequency domain. The parameters of the surface impedance model are usually fitted for sole conductors which enables an accurate modeling of the skin effect.

An alternative is presented in [97], where the system is modeled in two stages. In the first stage, the volumetric approach is applied by using a bundle of filaments and extracting the internal impedance for each detached conductor. In the second step, each conductor is modeled as a single volume cell while all mutual interactions are considered. At the same time, the actual DC resistance is replaced by the prior extracted internal impedance. This consequently leads to an accurate modeling of the skin effect while reducing the overall system size.

However, all of the presented techniques avoiding the full volume mesh lack of capturing the proximity effect caused by the mutual internal coupling effects of nearby conductors such as occurring in multi-turn coils.

Recently, a new surface PEEC formulation has been developed [98] which completely avoids a resolving of the interiors of the conductors while accounting for all physically relevant effects. However, this technique is plagued with low-frequency instabilities [23, 98] and is consequently not a good candidate for low- and medium-frequency IPT systems. In [44], another new integral equation based method is proposed which eliminates the volume currents by using a mathematical substitution. It is stated in the reference that the low frequency instabilities are overcome, thus obtaining a full spectrum method. However, this formulation is no longer compatible with PEEC and no network representation is obtained.

A different approach to reduce the number of elements needed in the volume based formulation is to apply specialized volume current basis functions [23]. These basis functions can account for the high current density at the boundaries of the conductors in a more precise manner compared to the standard piecewise-constant basis functions. The method also captures the full-spectrum frequency range. Moreover, it is able to account for the proximity effect by adapting the choice of the basis functions. This is achieved by using a fitting algorithm which estimates the general behavior of the current distribution by pre-solving some test cases with different conductor arrangements and frequencies.

Although the aforementioned approach with specialized basis functions could be an interesting alternative for the PEEC modeling of IPT systems, the standard volume discretization method is chosen in this work. The main motivation is the generality and the easy-to-implement character as well as the fact that a full volume discretization of a multi-turn PSC typically does not exceed a few thousand elements.

4.6.2 Subdivision of the Conductor's Cross Sections

As already motivated by the previous considerations, the frequency-dependent resistive and inductive behavior due to the skin and proximity effects is captured by a subdivision

4.6 Modeling of Skin and Proximity Effects

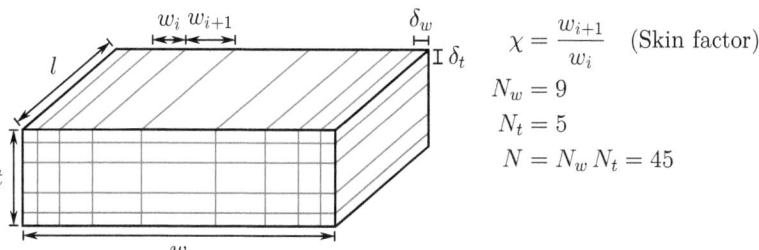

Figure 4.11: Volume discretization of a rectangular conductor with N non-equidistant segments. Towards the interior, each element is increased by the constant factor χ which reflects the decaying current distribution towards the interior of the conductor. The skin factor is typically chosen in between $2 \leq \chi \leq 3$.

of the cross sections of the straight conductors. A common starting point is to choose the width of the outermost volume cell to be less or equal to the half skin depth δ at the highest frequency of interest which is a tradeoff between accuracy and numerical effort. Contrary, a relatively coarse discretization of the interior of the conductor is sufficient since the current is generally more dense at the surface of the conductors. This motivates to set up a non-equidistant mesh which is coarser at the interior and finer at the boundary region. For the sake of simplicity, the width of the elements is increased towards the interior of the conductors by a constant factor χ which will be referred to as *skin factor* in the following. The influence of the skin factor is visualized in Figure 4.11 and has to be determined prior to the meshing of the structures. The skin factor typically ranges in between two and three which reflects a good tradeoff between the reduction of elements and the discretization error as has been verified in the case study in appendix C on page 177. It is shown in there that the introduced discretization error is of about 1 % for a broad parameter range.

A note should be made about the cross-sectional geometry of the volume cells. Although arbitrary cross sections of the basic PEEC current cells can be set up, rectangular cross sections are used throughout this work. This is because rectangular elements can account for arbitrary conductor cross sections without gaps. Additionally, a perfect approximation for conductors with rectangular cross section is enabled. Moreover, analytical integral solutions are available for the rectangular element coupling integrals. On the other hand, it should not be concealed that basic cells of triangle or curvilinear shape would be more efficient for approximating conductors with circular or arbitrary curved cross sections.

Rectangular Cross Section In Figure 4.11, the discretization of a rectangular conductor with $N = 45$ bricks is visualized. As stated before, an accurate modeling of

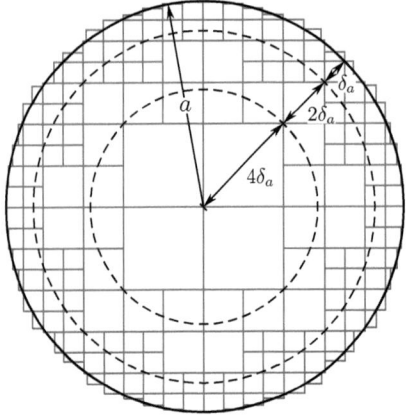

Figure 4.12: Discretization scheme of a cylindrical conductor with a set of rectangular elements. The meshing algorithm first divides the cross section into a set of circular disks, whereas the radial difference of two neighboring disks halves when going outwards. Subsequent, the side length of the square elements in each disk level is halved in order to account for the higher current density at the surface. In this case, the number of cells is 196 and $\delta_a = a/7$.

the skin effect demands the outermost segment size characterized by δ_w and δ_t to be comparable to the skin depth δ of (3.10). When introducing the skin factor χ according to Figure 4.11 with $\delta_{\max}$ being the maximum width of the outermost segment, the necessary number of subdivisions N_w in the w-direction can be computed according to [99, eq. (3.63)] as

$$N_w = \left\lceil \frac{2}{\ln(\chi)} \ln\left(\frac{(\chi-1)w + 2\delta_{\max}}{(\chi+1)\delta_{\max}}\right) + 1 \right\rceil, \quad \text{with} \quad w > 2\delta_{\max}. \tag{4.36}$$

In the above equation (4.36), $\lceil \cdot \rceil$ denotes to round up the argument to the next greater odd integer. It is always recommended to choose an odd number of subdivisions in order to avoid an unnecessary symmetrical discretization of the innermost area of the conductor. Typically, the maximum allowed segment width $\delta_{\max} = \delta/2$ is chosen as being half of the skin depth which is also referred to as the $\delta/2$-rule.

The above equation (4.36) allows in turn to compute the actual width of the outermost segment δ_w in a straightforward manner from which follows $\delta_w \leq \delta_{\max}$.

The number of segments N_t in t-direction and the corresponding δ_t can be computed accordingly to (4.36) by substituting w by t. For some applications it is further convenient to express the geometric mean width of the outermost segment defined as $\delta_{wt} = \sqrt{(\delta_w \delta_t)}$.

Circular Cross Section In Figure 4.12, the mesh of a conductor with a circular cross section of radius a by means of square elements is presented. Although the discretization of a circular disk via rectangular patches is suboptimal and discretization errors are introduced, the rectangular mesh is used for verifying the discretization method since an analytical reference solution is available for this type of conductor.

The meshing algorithm presented in Figure 4.12 works as follows: First, the cross section is divided into n circles, while the radial difference of two subsequent circles is halved each time from the center to the outward direction. The virtual circles then partition the cross section into n circular rings which describe areas of equal discretization levels. Second, the width of the outermost circular ring is defined as δ_a. This determines the discretization accuracy since the largest element in this area does not exceed δ_a. Furthermore, the algorithm checks if small elements can be added at the surface in order to better approximate the curved boundary. At the end, the algorithm checks in each circular level whether four square neighboring elements can be merged together in order to reduce the overall number of elements.

A convergence study of the internal impedance of the cylindrical conductor with circular cross section will be presented in section 6.1.2.

4.7 Modeling of Materials

This section is focused on the inclusion of linear dielectric and magnetic materials into the PEEC method. Some information about the general properties of such materials has already been presented in section 2.3. Compared to magnetic materials, dielectric materials have been paid more attention in the past decades since the PEEC method has originally been developed for simulating interconnection structures which are typically mounted on dielectric substrates. However, modeling of magnetic substrates is of interest for inductive applications because the magnetic field distribution can be influenced by magnetic materials. In recent years, intensified research towards this kind of materials has been emerged. For this reason, a few aspects about magnetic-material modeling will be reviewed in section 4.7.2 although the actual code implementation of this work does not support magnetic materials.

From a technical point of view, there exist two different fundamental concepts for the inclusion of dielectric and magnetic materials. In the first approach, the additional charges and currents of (2.30c) respective (2.30d) are modeled in the PEEC method by discretizing the material regions. Alternatively to that, the Green's functions (2.32) can be adapted in order to account for the differential equations which are valid for the actual media distribution. This approach is often used for stratified media such as multilayer PCB structures.

4.7.1 Dielectric Materials

The inclusion of arbitrarily shaped piecewise linear dielectrics for the full-wave PEEC models has first been proposed by Ruehli and Heeb [75, 76]. The basic concept of this approach will be repeated in the following.

Chapter 4 Partial Element Equivalent Circuit Method

First, the dielectric regions are discretized into a 1D, 2D or 3D grid according to the conducting regions as explained in section 4.5.1. On this grid, a set of unknown volume polarization currents $\vec{J}^{\mathrm{P}}$ according to (2.30d) and surface[11] polarization charges as in (2.30c) is assumed. These additional currents and charges are included in the total currents and charges of the MPIE formulation (2.34) which is then discretized according to the derivation of the beginning of this chapter. The deduction of the basic equations is equivalent for the dielectric cells and hence not repeated in detail. However, one main difference will be highlighted which occurs in the self-terms of the dielectric current cells w. r. t. the conductor current cells. More specifically, a capacitance rather than a resistance is obtained for the dielectric cells. This property can be highlighted when the different terms are compared next to each other.

In the basic MPIE formulation, the electric field strength inside the conductors is substituted by Ohm's law which is repeated on the left hand side of the following equation as

$$\vec{E} = \frac{1}{\kappa}\vec{J}, \qquad \vec{E} = \frac{1}{j\omega\,\varepsilon_0(\varepsilon_\mathrm{r}-1)}\vec{J}^{\mathrm{P}}. \qquad (4.37\mathrm{a})$$

The counterpart for dielectric regions on the right hand side is obtained by substituting (2.9) into (2.5c). Obviously, both relations account for the inherent properties of the materials. The different equations for conducting and dielectric cells from (4.37a) can be carried through the derivation of the PEEC method from section 4.1. Without repeating all steps, subsequent to the discretization and testing via the Galerkin method, the above system is transferred to the discrete formulation (cf. (4.8a))

$$\underline{U}_m = \frac{l_m}{\kappa_m A_m} \underline{I}_m, \qquad \underline{U}_n = \frac{1}{j\omega\,\varepsilon_0(\varepsilon_{\mathrm{r}n}-1)\,A_n/l_n} \underline{I}_n. \qquad (4.37\mathrm{b})$$

The fraction of the left expression in the above equation is interpreted as a partial resistance in (4.11) with the cell length l_m and cross section A_m. In accordance to that, the fraction of the right expression can be interpreted as a capacitive impedance term with the so-called *excess capacitance*, defined as [75, 76]

$$C_n^+ = \frac{\varepsilon_0(\varepsilon_{\mathrm{r}n}-1)\,A_n}{l_n}. \qquad (4.38)$$

This capacitance includes the relative permittivity of the dielectric cell. It can be extended to lossy materials by using the complex permittivity of (2.12) which can be accounted for in the network domain by applying an additional resistance connected

---

[11] As motivated in section 2.3.1, the polarization charges can be restricted to the surfaces of the piecewise homogeneous dielectric regions.

in parallel to the excess capacitance [86]. In addition, the model can also be used to include dispersive dielectrics [100].

Summing up, for dielectric cells the traditional resistance is replaced by the excess capacitance. The concept of partial inductance and coefficient of potential as well as the continuity equation in form of the KCL remain unchanged. As a consequence, the system size is enlarged due to the additional cells. Since every polarization current is mutually coupled with every conducting current while the same is true for the charges, the method can become cumbersome for large 3D dielectric regions.

An alternative to the discretization of the dielectric objects is to adapt the Green's function kernel in order to account for the media, which is often done in the context of multilayer structures. Here, no modeling of the polarization currents and charges is required. However, the drawback is the growing complexity of the partial element computations since the Green's function kernel becomes mathematically more complex. For further reading, the reader is referred to the literature, e.g. [22, 23, 101]. In this work, the general full-wave case with multilayer PCB structures is not pursued any further. Instead, a low-frequency approach will be discussed in the following.

Electrostatic Modeling As motivated at the end of section 2.4.1, the polarization currents may be neglected for LQS systems.[12] As a consequence, a quasi-static formulation without these currents is obtained as can be verified by (2.30d). According to the above, two methods for including the polarization charges exist, either by additionally discretizing these charges or, alternatively, by adapting the Green's function.

The first approach is also known as Equivalent Charge Formulation (ECF) [45, 102] in which the surface polarization charges ϱ^P of (2.30c) are discretized. The material properties are accounted for by relating the normal components of the electric flux density at the interfaces [45]. Thus, a higher order electrostatic system is obtained. One possibility of integrating the extended system into the standard PEEC formulation is to compress the obtained system which leads to an adapted matrix of the panel coefficients of potential in (4.15b). This allows for simulating the interconnection system as if the dielectric components would not be present, but using adapted potential and capacitance matrices in which the influence of the dielectrics is included.

The alternative, especially for stratified media such as multilayer PCBs, is to adapt the Green's function as before. The general procedure is explained, for instance, in [103]. The Green's function for a dielectric substrate can be regarded as a method of images for a point charge. In the case of a single dielectric transition, a single mirror charge is obtained while in case of multiple transitions, the Green's function is represented as an infinite series [73].

In case of a two-layer substrate, a closed-form solution is presented in [46]. In the reference, an infinite series representation accounts for the multiple reflections of the

---

[12]In [75, p. 978] it is stated that this method produces a good approximation to the full-wave method.

image charges. When this Green's function is substituted in the coefficient-of-potential expression (4.14a), the analytical expressions for rectangular patches as presented in appendix A.4 on page 164 can be maintained since the integration and summation can be interchanged [73]. As a matter of fact, the infinite series is truncated after n mirror charges which introduces an additional error and increases the effort to compute each coefficient of potential by the factor of n.

The actual implementation of the PEEC solver is based on this formulation. The expression of the Green's function in this case is detailed in appendix A.5 on page 166.

4.7.2 Magnetic Materials

The common approach to include linear magnetic materials into the PEEC method is to model the magnetization current density $\vec{J}^M$ of (2.30d). Although this current density is generally of volumetric nature, even for piecewise homogeneous materials (2.15), it is typically modeled as a surface current only [77]. This has been motivated at the end of section 2.4.1. Hence, the surfaces of the material blocks are discretized and unknown magnetization currents are assumed providing new voltage sources in the basic PEEC cells.

The magnetization currents are typically not interpreted as equivalent circuit cells. Instead, they are accounted for by the constitutive relations. This results in a further system of equations [77] which has either to be solved together with the MNA system or, alternatively, can be used to adapt the total inductance matrix [100]. Other contributions which investigate on the integration of magnetic materials into the PEEC models are [104, 105, 106].

4.8 Acceleration Techniques

In the last section of this chapter it is briefly argued why the classical acceleration techniques used in PEEC models are not or not simply applicable to the IPT antenna systems composed of multi-turn PSCs. Basically, all methods have in common to avoid the dense matrix fills of the inductance and coefficient of potential matrices since the time and storage requirements grow with order $O(N^2)$.

The *reluctance-based* method of [107, 108] obtains a diagonal dominant new network-element reluctance matrix which is defined as being the inverse of the inductance matrix. The matrix of partial reluctances can be shown to be more locally or diagonal dominant compared to the inductance matrix [107]. Consequently, the errors introduced by neglecting couplings from distant current cells are smaller compared to neglecting the corresponding terms in the inductance matrix.

Another class of acceleration techniques is given by the Fast Multipole Method (FMM) which is detailed in case of the PEEC method in [109], for example. The FMM also avoids the complete matrix fill. As a prerequisite, it is based on an iterative solution of the linear system of equations. The element couplings are partitioned into weak coupling elements of more distant cells on the one hand and neighbour interactions of elements located in close proximity on the other hand. Only the interactions of these cells are explicitly computed and stored. The couplings of all other elements are evaluated on-the-fly in each iteration. By reformulating the Green's function into a multipole expansion [31], the interactions are realized by so-called group centers to which the cells are assigned to.

Although the reluctance-based method as well as the FMM may be advantageous for various applications, the applicability to skin- and proximity-effect problems such as occurring in multi-turn PSCs is limited. This is due to the high aspect ratios of the current cells and the close proximity at the same time (s. Figure 4.6). Thus, in the reluctance-based method, the negligible element couplings are not significant while in the fast multipole method, group interactions are of minor importance if not vanishing.

Chapter 4 Partial Element Equivalent Circuit Method

Chapter 5
Sensitivity Analysis

During the design and optimization processes of computer aided engineering, it is favorable to obtain information about the influence of parameter variations on the system behavior. The parameters of interest will be referred to as design parameters in the following and may be shape coefficients for instance. In case of Inductive Power Transfer (IPT) system design parameter tolerances which occur during a usually non-perfect manufacturing process can influence the overall system behavior in terms of efficiency, quality factors and the Self-Resonant Frequency (SRF). In order to quantify the influence of pa-

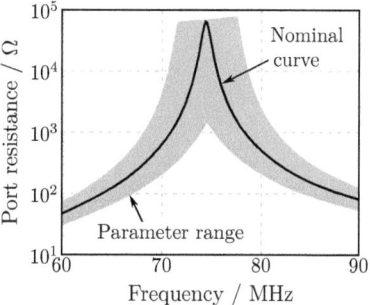

Figure 5.1: Sensitivity Example

rameter changes, it is convenient to use a gradient representation in the parameter space which corresponds to a linearization around the nominal working point. This allows a prediction of the system behavior for small parameter changes.

The sensitivity analysis is a general approach to obtain such derivative information which is needed for the gradient representation. If the so-called adjoint sensitivity analysis is applied to EM field solvers, the mentioned gradient information is obtained by evaluating the system twice, independently of the number of design parameters. At the same time, the adjoint sensitivity analysis demands a computation of the derivatives of the system matrix entries w.r.t. the design parameters. This may become one of the major cost factors in the overall simulation time. In the context of the PEEC method, the system matrix consists of the partial network elements as has been widely discussed in the previous chapter. In this work, the demanded derivatives w.r.t. the design parameters will either be approximated by Finite Difference (FD) approximations or computed exactly by means of closed-form expressions which are available for certain geometrical shapes of the basic cells.

The chapter is structured as follows: First, the fundamental equations needed for the adjoint sensitivity analysis are presented. Followed by that, the sensitivity analysis is applied to the MQS-PEEC method, especially in terms of an efficient derivative computation for skin- and proximity-effect problems. Numerical results will be examined for two different examples, the single rectangular conductor and a multi-turn PSC in section 6.1.5 and section 6.2.3 on page 142, respectively.

5.1 Adjoint Based Method

The following considerations are based on a linear system of equations of the general form

$$\mathbf{A}\mathbf{x} = \mathbf{b}, \tag{5.1}$$

where $\mathbf{A}$ denotes the $N \times N$ complex valued system matrix, $\mathbf{x} = [x_1 \ldots x_N]^T$ the $N \times 1$ vector of state variables and $\mathbf{b} = [b_1 \ldots b_N]^T$ the $N \times 1$ excitation vector with N being the number of elements.[1] In the following, it is assumed that the system matrix elements as well as the entries of the excitation vector may depend on N_d design parameters $\mathbf{p} = [p_1 \ldots p_n \ldots p_{N_d}]^T$. The n-th parameter p_n typically characterizes material properties or geometrical shape parameters such as the thickness of a conductor. As a consequence, the state variables also depend on the design parameters.

In order to quantify the dependence of the system behavior on the parameters, the common way is to differentiate the vector of state variables w.r.t. p_n. This allows a linearization of the system behavior in form of a Taylor series containing the linear terms. Applying the method to (5.1) and using the product rule of differentiation, the following equation is obtained

$$\frac{\partial \mathbf{A}}{\partial p_n}\mathbf{x} + \mathbf{A}\frac{\partial \mathbf{x}}{\partial p_n} = \frac{\partial \mathbf{b}}{\partial p_n}, \tag{5.2a}$$

from which the dependence off all state variables w.r.t. the design parameter p_n according to

$$\frac{\partial \mathbf{x}}{\partial p_n} = \mathbf{A}^{-1}\left[\frac{\partial \mathbf{b}}{\partial p_n} - \frac{\partial \mathbf{A}}{\partial p_n}\mathbf{x}\right] \tag{5.2b}$$

can be derived. Often, the sensitivity of a deduced quantity such as an impedance or scattering parameters is desired rather than the sensitivity information of the entire vector of state variables (5.2b). For this reason, an arbitrary response or objective function $f(\mathbf{p}, \mathbf{x}(\mathbf{p}))$ is introduced which generally may depend on the state variables as well as the parameters $\mathbf{p}$ explicitly. The differentiation of the objective function w.r.t. the design parameter p_n can be expressed by means of the chain rule in case of multiple variables as

$$\frac{\partial f}{\partial p_n} = \frac{\partial^e f}{\partial p_n} + \frac{\partial f}{\partial x_1}\frac{\partial x_1}{\partial p_n} + \ldots + \frac{\partial f}{\partial x_N}\frac{\partial x_N}{\partial p_n} = \frac{\partial^e f}{\partial p_n} + \nabla_\mathbf{x} f \frac{\partial \mathbf{x}}{\partial p_n}, \tag{5.3}$$

---
[1] As a consequence of this standardized notation, units are not accounted for in this section although the concrete realization of (5.1) may exhibit physical units.

5.1 Adjoint Based Method

whereas the possible explicit dependence of $\underline{f}$ on p_n is represented by $\partial^e \underline{f}/\partial p_n$. The gradient operator $\nabla_{\mathbf{x}} = [\partial/\partial x_1 \cdots \partial/\partial x_N]$ accounts for the differentiation w.r.t. $\mathbf{x}$ and is interpreted as a row operator. Consequently, the expression $\nabla_{\mathbf{x}} \underline{f}$ indicates how the objective function $\underline{f}$ is influenced by $\mathbf{x}$.

As an example, the MNA formulation of a one-port network in (4.21) is considered. The port voltage $\underline{U}_{\mathrm{port}}$ is defined as the objective function which linearly depends on the state vector $[\underline{\mathbf{i}}_{\mathrm{b}}, -\underline{\boldsymbol{\varphi}}_{\mathrm{n}}]^{\mathrm{T}}$ as can be verified by (4.21b). In this case, $\nabla_{\mathbf{x}} \underline{f} = -[\mathbf{0}, \mathbf{K}^{\mathrm{T}}]$ is a row vector, purely composed of ± 1 and 0 entries according to (4.20).

The above expression (5.3) allows for setting up the response sensitivity equation by substituting (5.2b) into (5.3) which leads to[2]

$$\frac{\partial \underline{f}}{\partial p_n} = \frac{\partial^e \underline{f}}{\partial p_n} + \nabla_{\mathbf{x}} \underline{f} \, \mathbf{A}^{-1} \left[\frac{\partial \underline{\mathbf{b}}}{\partial p_n} - \frac{\partial \mathbf{A}}{\partial p_n} \underline{\mathbf{x}} \right]. \tag{5.4}$$

The product $\nabla_{\mathbf{x}} \underline{f} \, \mathbf{A}^{-1}$ of the above equation, which is independent of the design parameters $\mathbf{p}$, can be combined to a single vector as

$$\hat{\underline{\mathbf{x}}}^{\mathrm{T}} = \nabla_{\mathbf{x}} \underline{f} \, \mathbf{A}^{-1}. \tag{5.5a}$$

The interpretation of $\hat{\underline{\mathbf{x}}}^{\mathrm{T}}$ can be clarified when (5.5a) is first transposed and afterwards left-multiplied by $\mathbf{A}^{\mathrm{T}}$, resulting in the *adjoint system*[3]

$$\mathbf{A}^{\mathrm{T}} \hat{\underline{\mathbf{x}}} = [\nabla_{\mathbf{x}} \underline{f}]^{\mathrm{T}}. \tag{5.5b}$$

As a consequence, the vector $\hat{\underline{\mathbf{x}}}$ is referred to as the adjoint-variable vector [110] since it is the vector of state variables of the new system of equations. When this system (5.5b) is compared with the original one (5.1), two differences become obvious, the transposed system matrix on the one side and the different excitation vector on the other.

The vector $\hat{\underline{\mathbf{x}}}$ is obtained by the solution of the adjoint system (5.5b) and does not depend on the design parameters. Substituting the solution of the adjoint system (5.5) into (5.4), the sensitivity of $\underline{f}$ w.r.t. p_n is obtained as

$$\frac{\partial \underline{f}}{\partial p_n} = \frac{\partial^e \underline{f}}{\partial p_n} + \hat{\underline{\mathbf{x}}}^{\mathrm{T}} \left[\frac{\partial \underline{\mathbf{b}}}{\partial p_n} - \frac{\partial \mathbf{A}}{\partial p_n} \underline{\mathbf{x}} \right]. \tag{5.6a}$$

Often, the objective function does not explicitly depend on the design parameters. If the same is true for the excitation vector $\underline{\mathbf{b}}$, the above system can be simplified to

$$\frac{\partial \underline{f}}{\partial p_n} = -\hat{\underline{\mathbf{x}}}^{\mathrm{T}} \frac{\partial \mathbf{A}}{\partial p_n} \underline{\mathbf{x}}. \tag{5.6b}$$

---

[2] As explained in [110], it is also possible to rewrite (5.4) as a variant which accounts for all design parameters at the same time.

[3] The terminology adjoint system originates from the Tellegen's theorem [111], in which two networks are compared, the original and the adjoint (transposed) system.

Chapter 5 Sensitivity Analysis

Independent of the number of design parameters, just two systems have to be solved, the original system with the solution $\mathbf{x}$ and the adjoint system with the solution $\hat{\mathbf{x}}$. If an LU decomposition is performed to solve the original system, the overhead for the adjoint system is one forward and backward substitution only (cf. [112]). Furthermore, if the system matrix is symmetrical as in the case of the MNA system[4] (4.19) and the excitation vector of the adjoint system is identical to the original system as in the above example,[5] the solution of the adjoint system is identical to the original system. In this case it can be reused without any computational overhead.[6]

For evaluating (5.6), the various derivatives of the system matrix $\partial \mathbf{A}/\partial p_n$ have to be computed element by element. This can be carried out either analytically or numerically. A numerical realization in terms of a finite difference approximation can be represented as [114]

$$\frac{\partial \mathbf{A}}{\partial p_n} = \begin{bmatrix} \frac{\partial}{\partial p_n} A_{11} & \cdots & \frac{\partial}{\partial p_n} A_{1N} \\ \vdots & \ddots & \vdots \\ \frac{\partial}{\partial p_n} A_{N1} & \cdots & \frac{\partial}{\partial p_n} A_{NN} \end{bmatrix} \approx \frac{\Delta \mathbf{A}}{\Delta p_n} = \frac{\mathbf{A}_n - \mathbf{A}}{\Delta p_n}, \tag{5.7}$$

in which an additional matrix fill $\mathbf{A}_n$ must be set up for the perturbed parameter p_n. It should be mentioned that the sizes of the matrices $\mathbf{A}_n$ and $\mathbf{A}$ must be identically. As an example, a deformation of the mesh is allowed but the topology and size of the system must be maintained. Since the system matrix is dense in integral equation based formulations, this additional matrix fill may cause the main part of the overall time needed for the sensitivity analysis. Moreover, it must be repeated for each design parameter p_n.

When concretizing the system matrix to the MNA formulation (4.19) as before and assuming an independence of the nodal connectivity matrix of the design parameters, the matrices of the partial elements have to be differentiated w.r.t. the design parameters. When inspecting the partial network element definitions of section 4.2, the equations mainly depend on the materials and the geometry. A method for the computation of the exact geometrical derivatives of the partial inductances is presented in [115]. For completeness, the expressions are repeated in appendix A.2 on page 162. Although the use of exact derivatives does not significantly speed up the simulation time compared to finite difference approximations, a considerable advantage is obtained. This is the usability of the results as reference values, for instance, when determining an optimum step size needed for FD approximations.

Summing up, the equations required for the adjoint sensitivity analysis are given by the original system (5.1) together with the adjoint system (5.5b) and the sensitivity

---

[4]It is assumed that the standard PEEC formulation is used where a Galerkin procedure is applied to obtain the partial element matrices.
[5]The only difference is the negative sign.
[6]See also [113] in the case of MoM systems.

equation (5.6). A specific example is presented in section 6.1.5 where the skin effect impedance of a single conductor is differentiated w. r. t. the conductor width.

5.2 Inner-Layer Concept for Skin-Effect Sensitivities

In this section, different approaches are presented which reduce the effort of the additional matrix setups needed for computing the derivatives in the adjoint sensitivity analysis. Especially the skin-effect modeling is focused on as this effect plays an important role in IPT systems which has already been motivated in chapter 3. When transferring the general adjoint sensitivity equation (5.6) to the PEEC system, the matrix of partial inductances[7] $\mathbf{L}$ has to be differentiated w. r. t. the geometrical parameters as well as the underlying material properties.

Due to the fact that this computation needs to be repeated for each design parameter, it is desirable that a geometrical parameter perturbation affects only the neighboring segments in a local sense. Hereby, only a sub-region of the matrix entries needs to be recomputed. Because of the dense characteristic of the inductance matrix in which usually each element is coupled with all other elements, the locality is generally difficult to obtain.

Consider the single-conductor example according to Figure 5.2a, in which the width w is regarded as being the design parameter p_n. As can be seen in Figure 5.2b, this example reflects the worst-case scenario since the perturbation of the width by Δw affects all current cells. In this case, the system matrix density is equal to one. Thus, the computation of

$$\frac{\partial \mathbf{L}}{\partial w} = \begin{bmatrix} \frac{\partial L_{11}}{\partial w} & \cdots & \frac{\partial L_{1N}}{\partial w} \\ \vdots & \ddots & \vdots \\ \frac{\partial L_{N1}}{\partial w} & \cdots & \frac{\partial L_{NN}}{\partial w} \end{bmatrix} \tag{5.8}$$

requires the calculation of all matrix-element derivatives. This is at least as expensive as building the original matrix and consequently becomes cumbersome for large systems with multiple design parameters. It should be pointed out that due to the symmetric property of the inductance matrix, only the upper right part of the matrix including the main diagonal self-inductance terms has to be filled. This leads to the total number of entries of $1/2\,(N^2 + N)$ according to (Fig. 5.2-2).

In order to reduce the number of element computations, one approach is to perturb the elements at the boundary only, for instance [116] in the case of MoM applications. This method is referred to as Boundary-Layer Concept (BLC) in the following. As can

---

[7]The coefficients of potential are not discussed in this section since on the one side, skin-effect problems can be regarded under the MQS assumption. On the other side, the extension to the coefficients of potential is straightforward and generally more simple compared to partial inductances. This is because only double-surface integrals (4.14) need to be considered instead of double-volume integrals (4.12).

Chapter 5 Sensitivity Analysis

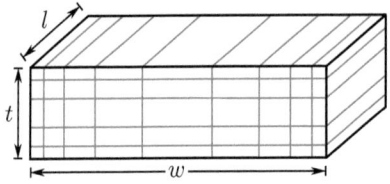

Number of current elements:

$$N = N_w N_t = 9 \cdot 5 = 45 \quad \text{(Fig. 5.2-1)}$$

(a) Original conductor

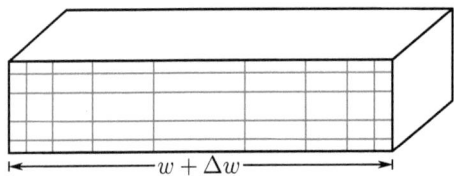

Number of mutual element couplings:

$$\frac{1}{2}(N^2 + N) \approx \frac{1}{2}N^2 \quad \text{(Fig. 5.2-2)}$$

(b) Uniform Perturbation

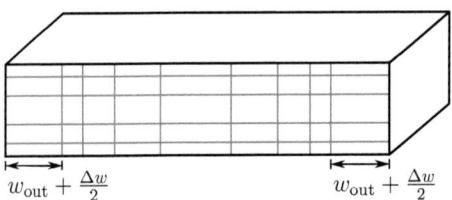

Number of mutual element couplings:

$$2N_t^2(N_w - 1) + N_t < \frac{2}{N_w}N^2$$
$$\text{(Fig. 5.2-3)}$$

(c) Boundary-Layer Concept, **Speedup min. $N_w/4$**

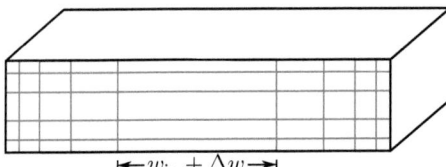

Number of mutual element couplings:

$$\frac{1}{4}N_t^2\left[(N_w + 1)^2 - 2\right] + \frac{1}{2}N_t > \frac{1}{4}N^2$$
$$\text{(Fig. 5.2-4)}$$

(d) Inner-Layer Concept, **Speedup max. 2**

Figure 5.2: Different approaches for perturbing the current elements when varying the width of a rectangular conductor as presented in (a). The standard approach (b) is to perturb all segments in the same manner. A symmetric alternative which saves most computational time is to perturb the outermost segments only (c). However, this method is not adequate for the skin-effect modeling since the change of current density is extremely high at the boundaries. A method which perturbs the inner segments only is presented in (d). This approach is more appropriate for skin-effect problems due to the usually low current density in the interior of the conductor.

5.2 Inner-Layer Concept for Skin-Effect Sensitivities

be verified by Figure 5.2c, this approach significantly reduces the number of element computations by a factor of more than $N_w/4$ with N_w being the number of subdivisions of the conductor in width direction. This is because the mutual couplings of the interior elements are not influenced by the perturbation. Consequently, some sub-regions of (5.8) are zero. Although the BLC might be appropriate for various applications, it is difficult to be applied to skin-effect problems since the current concentrates in the region of the boundary of the conductor elements. Moreover, a sole perturbation of the outermost elements hardly reflects the physical behavior. In Figure 6.12 on page 121, the convergence of the BLC for a specific test setup is analyzed. From the setup it becomes obvious that the obtained errors are not satisfactory.

In order to overcome the mentioned drawback of low accuracy, a method called Inner-Layer Concept (ILC) is introduced in Figure 5.2d in which only the inner segments of the conductor are perturbed by Δw. The motivation for this technique is given by the fact that the estimated current density is relatively low at the innermost segments. Consequently, the introduced errors do not contribute too much to the overall system behavior. Since the left and right element blocks of the perturbed innermost segments are shifted during the perturbation, the computation of the elements is only reduced according to (Fig. 5.2-4) which corresponds to a speedup of approximately two when compared with the uniform perturbation.

Comparisons between the three different methods displayed in Figure 5.2 will be presented in section 6.1.5 for a single conductor example. All partial-element derivatives are computed there via analytical expressions in order to exactly compare the methods w. r. t. effort and accuracy.

Chapter 5 Sensitivity Analysis

Chapter 6
Simulation Results and Measurements

In this chapter, the analysis and numerical modeling of IPT antenna systems is tested with a typical antenna setup that is often employed in RFID systems. Numerical results will be performed with the developed PEEC code which is optimized for IPT systems by using numerous settings according to the description of chapter 4. Before analyzing a mutually coupled coil system, a single conductor as well as a single PSC are addressed in order to structure the analysis in a reasonable order and to successively use the findings and results of the initial investigations for the subsequent ones. In order to verify and validate the applicability of the used approach, the PEEC results will be compared with results of 2D- and 3D-FEM solvers as well as analytical results and measurements.

6.1 Cylindrical Conductor

In the first section, the current distribution and the AC impedance of a single cylindrical conductor are analyzed under MQS assumptions. This helps to verify the PEEC method as well as the sensitivity analysis and to find adequate mesh settings for an accurate modeling of skin and proximity effects occurring in PSCs at high frequencies. Two different cross sections are presented, the circular and the rectangular ones whereas the conductor with circular cross section is only regarded in order to compare the PEEC method with analytical results. For the rectangular shape, comparisons with analytical approximations as well as results of a 2D-FEM solver will be presented.

6.1.1 PEEC Solver Settings

The test setup for computing the AC impedance of the cylindrical conductor with the 2D-PEEC and MQS-PEEC methods is shown in Figure 6.1. A set of rectangular basic cells is set up to model the arbitrary cross section of the conductor. All partial resistances and inductances are computed, thus allowing for a frequency-dependent modeling of the current distribution which is represented by the vector of branch currents $\mathbf{i}_b$ through

Chapter 6 Simulation Results and Measurements

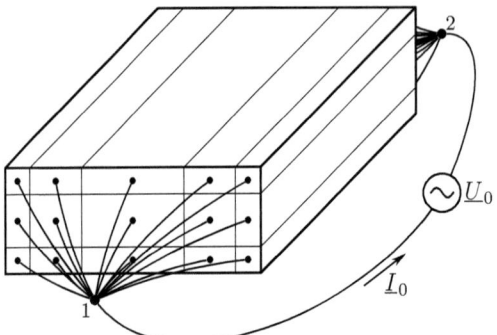

Figure 6.1: Test setup for analyzing the internal MQS impedance $Z_0 = U_0/I_0$ of a long cylindrical conductor via the PEEC method, visualized with an exemplary rectangular cross section. All current cells are connected in parallel between the two nodes whereas one node is defined as the reference node. The influence of the thin connecting wires is not modeled.

the cells. By using the nodal network theory and defining one of the two nodes as the reference node, the nodal connectivity matrix $\mathbf{B}$ from (4.9) becomes a $N_\mathrm{b} \times 1$ vector with all entries being identical to $+1$. This is because all elements are connected in parallel. Thus, the system (4.25) reduces with $\mathbf{u}_\mathrm{s} = 0$, $\mathbf{i}_\mathrm{s} = I_0$ and $\boldsymbol{\varphi}_\mathrm{n} = U_0$ to

$$(\mathbf{R} + j\omega\,\mathbf{L})\,\mathbf{i}_\mathrm{b} = \mathbf{B}\,U_0, \tag{6.1a}$$

$$I_0 = \mathbf{B}^\mathrm{T}\mathbf{i}_\mathrm{b}, \tag{6.1b}$$

where U_0 acts as the excitation of the system. The branch current vector $\mathbf{i}_\mathrm{b}$ is computed by the initial solution of (6.1a) which in turn provides the total current by (6.1b). This allows for defining the impedance

$$Z_0 = \frac{U_0}{I_0} \tag{6.2a}$$

as the objective function. Afterwards, the adjoint sensitivity equation is expressed as [115]

$$\frac{\partial Z_0}{\partial p_n} = \frac{1}{I_0^2}\mathbf{i}_\mathrm{b}^\mathrm{T}\frac{\partial(\mathbf{R} + j\omega\,\mathbf{L})}{\partial p_n}\mathbf{i}_\mathrm{b}, \tag{6.2b}$$

which is a concrete form of (5.6b). The design parameter p_n may again be a shape coefficient of the conductor.

6.1.2 Circular Cross Section and Infinite Length

As already stated in the introductory words of this section, the conductor with a circular cross section is used to verify the 2D-PEEC approach since analytical expressions exist for the MQS current distribution as well as the internal impedance. The geometrical

6.1 Cylindrical Conductor

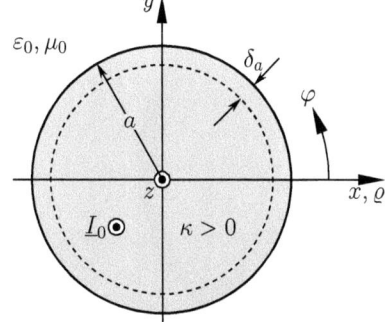

Table 6.1: Parameters of the test setup

$a = 1$ mm	$\kappa = 58 \; 10^6$ S/m

Figure 6.2: Circular cross section of the infinitely long cylindrical conductor with the radius a and the conductivity κ. According to Figure 4.12, the width of the outermost circular mesh ring δ_a is a measure for the discretization level.

dimensions are presented in Figure 6.2 with a being the radius of the conductor and κ the homogeneous conductivity. The total current through the conductor is specified by $\underline{I}_0$. For long conductors where the cross section is small compared to the length l, edge effects can be neglected and the originally 3D application can be simplified to a 2D problem.

Analytical Solution The frequency-dependent current distribution of the infinitely long circular conductor according to Figure 6.2 can be expressed under the MQS assumption in closed form as (cf. [50])

$$\underline{J}_z(\varrho,\omega) = \frac{\underline{I}_0 \, \underline{p}}{2\pi a} \frac{I_0(\underline{p}\varrho)}{I_1(\underline{p}a)}, \qquad J_z(\omega=0) = \frac{I_0}{\pi a^2}, \qquad (6.3a)$$

with $\underline{p} = (1+j)\sqrt{\omega\mu_0\kappa/2}$ and I_0 and I_1 being the modified Bessel functions of first kind and order zero and one, respectively. This in turn allows the expression of the per-unit-length resistance R' and the internal inductance L'_{int} as [50]

$$L'_{\text{int}}(\omega) = \text{Im}\left\{\frac{\underline{p}}{2\pi\kappa a}\frac{I_0(\underline{p}a)}{I_1(\underline{p}a)}\right\}\frac{1}{\omega}, \qquad L'_{\text{int}}(\omega=0) = \frac{\mu_0}{8\pi} = 50\,\text{nH m}^{-1}, \qquad (6.3b)$$

$$R'(\omega) = \text{Re}\left\{\frac{\underline{p}}{2\pi\kappa a}\frac{I_0(\underline{p}a)}{I_1(\underline{p}a)}\right\}, \qquad R'(\omega=0) = \frac{1}{\kappa\pi a^2}. \qquad (6.3c)$$

It should be noted that the external inductance cannot be expressed explicitly since the overall magnetic energy of (2.42) is infinite for infinitely long conductors. For this reason, only the internal inductance is considered for the following 2D applications. It is also worth mentioning that the internal inductance does not depend on the radius of the conductor and amounts exactly $50\,\text{nH m}^{-1}$ at the DC limit.

Chapter 6 Simulation Results and Measurements

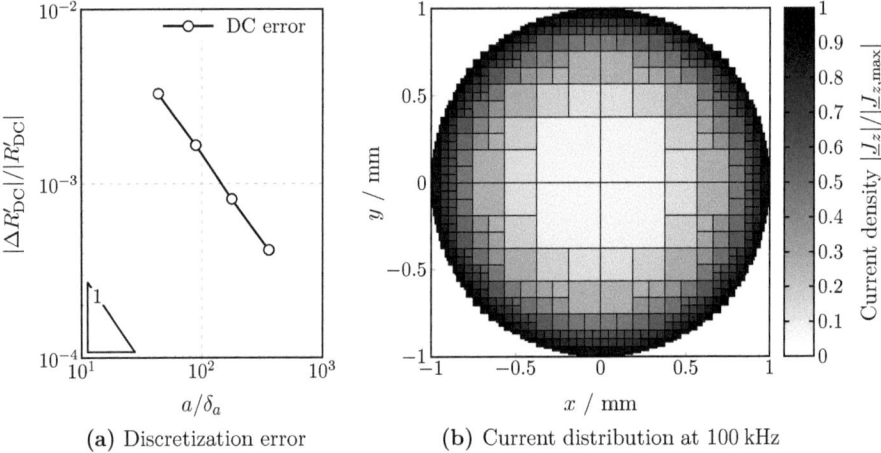

(a) Discretization error   (b) Current distribution at 100 kHz

Figure 6.3: (a) Convergence analysis of the DC-resistance error for decreasing δ_a. The error can also be regarded as the discretization error of the circle by means of rectangular cells. (b) Exemplary current distribution of the test conductor of Table 6.1 at $f = 100$ kHz. The outermost discretization level has a width of $\delta_a = 66.7$ μm leading to 552 elements.

2D-PEEC Model In order to validate the 2D-PEEC results and to test the implemented code in terms of meshing, mutual element interactions and system solving, the described meshing algorithm displayed in Figure 4.12 is tested with an exemplary conductor with the parameter settings chosen according to Table 6.1. In Figure 6.3a, the DC error of the PEEC model is visualized for an increasing mesh density. Due to the homogeneous current distribution at DC, the error corresponds to the discretization error caused by the approximation of the curved boundary by means of rectangular cells. If the ratio of the outermost-ring width by the radius is $\delta_a/a \leq 1/200$, the discretization error is well below 0.1 %.

Next, the analytically computed current density of (6.3a) is compared with the discretized counterpart obtained by the 2D-PEEC model. An exemplary current distribution at $f = 100$ kHz is visualized in Figure 6.3b. Due to the skin effect, the current concentrates at the outermost region of the conductor which justifies the relatively coarse mesh at the interior of the conductor. In order to better compare the 2D-PEEC current distribution with the exact one, a 1D plot on a cutting line is performed in Figure 6.4 for two different frequencies and discretization levels. The cutting line is located at $\varphi = 45°$ since the meshing algorithm according to Figure 4.12 uses the mesh cells on this line as the worst-case cells. It becomes obvious that the current distribution is approximated by the PEEC method in a stair-case manner. The essential point when comparing the

6.1 Cylindrical Conductor

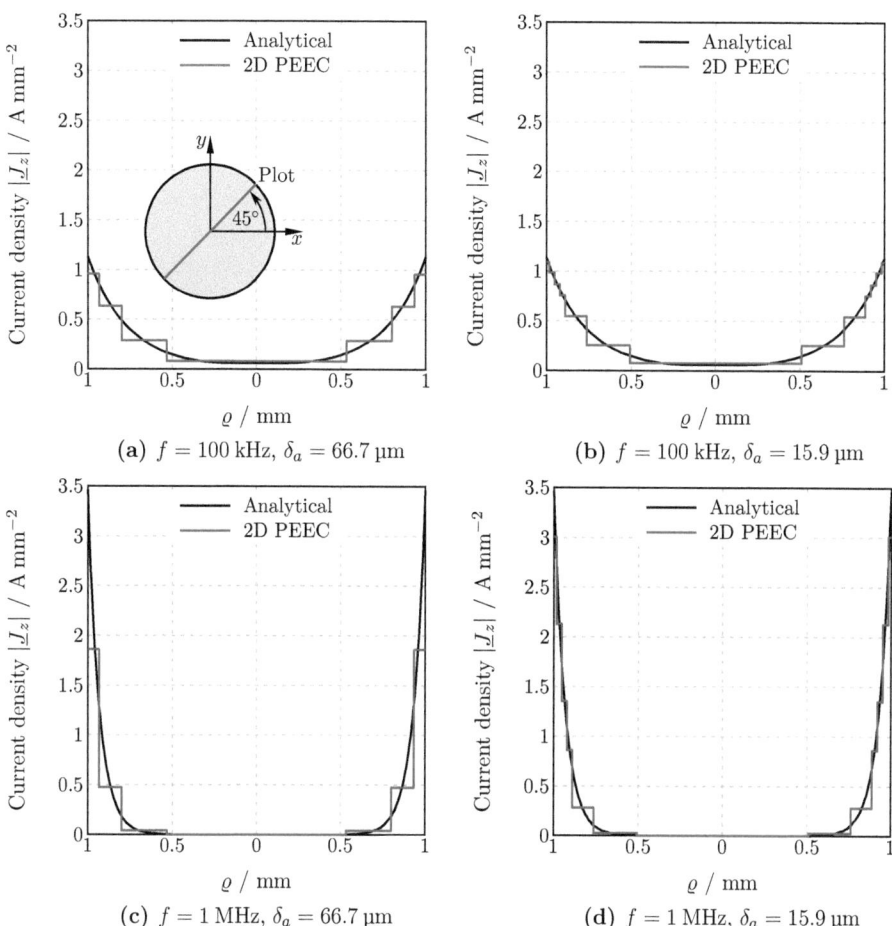

Figure 6.4: Current distribution of the test conductor of Table 6.1 for two different frequencies and discretization levels. The abscissa of the four plots is located on a cutting line at $\varphi = 45°$ as can be verified by the repeated outline of the conductor. In order to compare the current density at different frequencies, the total current is chosen to be 1 A in all cases. Whereas a coarse discretization (left figures) might be accurate enough for approximating the current distribution at low frequencies (top figures), it is not able to reflect the extremely high current density at the boundary occurring at high frequencies (bottom figures). In the right figures, a drawback of the meshing algorithm becomes visible. It is the fact that by increasing the discretization level, only the outermost area of the conductor is refined. Thus, the largest errors occur in the inner region of the conductor. This effect can also be seen in Figure 6.5 for medium frequencies.

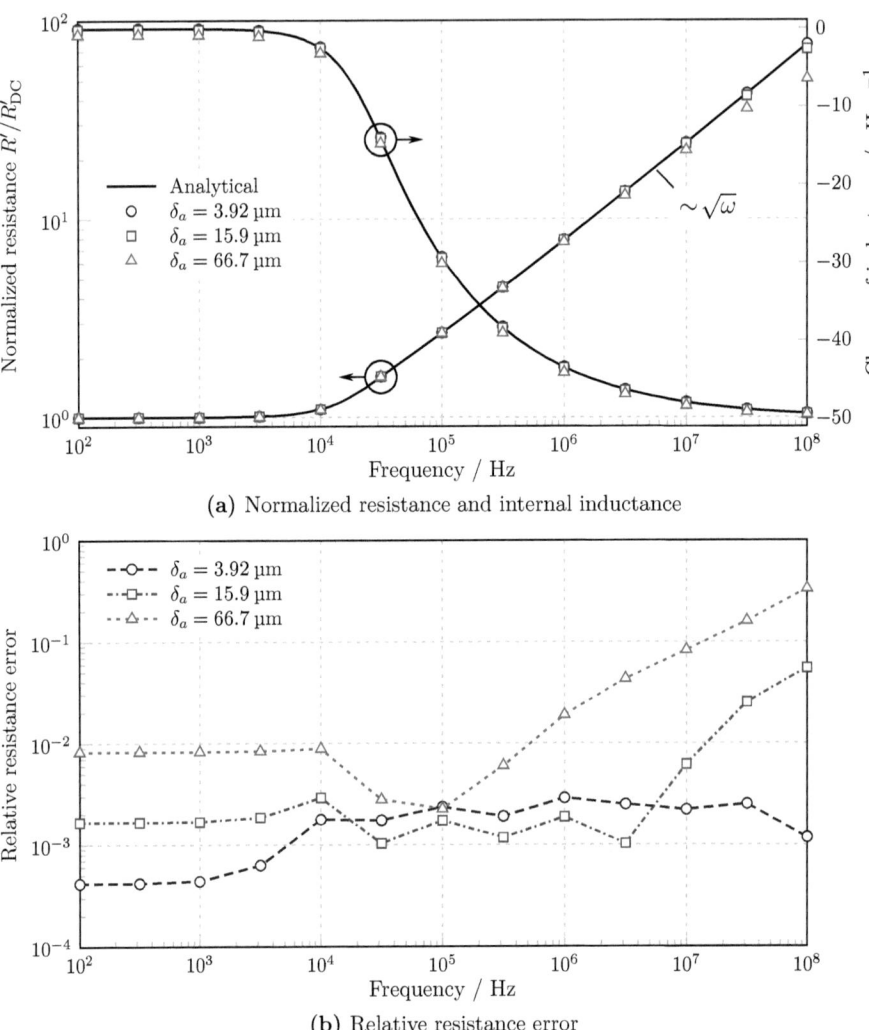

Figure 6.5: Resistance and inductance comparison for different discretization levels of the test conductor of Table 6.1. For low frequencies, the error behaves equivalently to the discretization error (s. Figure 6.3a). For medium frequencies, the error is dominated by the coarse discretization at the inner region of the conductor. This is relatively independent of the discretization levels. For high frequencies, the current crowds at the surface where the error is smallest for the highest resolution.

6.1 Cylindrical Conductor

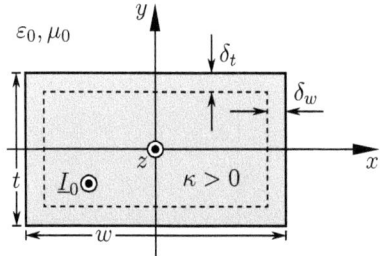

Table 6.2: Parameters of the test setup

$w = 1$ mm	$t = 0.1$ mm
$\kappa = 58 \; 10^6$ S/m	$l \to \infty$, $l = 50$ mm

Figure 6.6: Rectangular cross section of the infinitely long cylindrical conductor with the width w, thickness t and conductivity κ. The widths of the outermost segments δ_w and δ_t determine the discretization level.

different figures is the fact that a coarse discretization might be accurate enough for low frequencies while higher frequencies demand a finer mesh at the region of the boundary.

This statement is quantified more precisely in Figure 6.5 where the resistance and internal inductance are visualized over a broad frequency range for different discretization settings. For low frequencies, the error is mainly dominated by the discretization error determined by δ_a. For medium frequencies, the error is comparable for all different discretization levels since it is dominated by the coarse mesh at the interior of the conductor. For high frequencies, only the discretization with a mesh size comparable to the skin depth obtain acceptable errors.

Summing up, the 2D-PEEC method adequately accounting for the frequency-dependent internal effects of long conductors at the MQS assumption. However, the presented mesh with rectangular cells is not efficient because loads of elements are required in order to approximate the circular shape. Additionally, the volume discretization becomes inefficient for high frequencies where small elements are required at the surface of the conductor.

6.1.3 Rectangular Cross Section and Infinite Length

In this section, an infinitely long conductor with a rectangular cross section as sketched in Figure 6.6 is analyzed. In contrast to the circular cross section, the geometry is perfectly approximated by means of brick shaped elements. Due to the ease of producibility, conductors with rectangular cross sections are often used in various applications. On the other side, analytical reference solutions for computing the frequency-dependent current distribution or the internal impedance are difficult if not impossible to obtain. For this reason, the 2D-PEEC will be compared with a 2D-FEM solver as well as analytical approximation expressions. Results will be carried out for an exemplary test conductor with the parameter settings according to Table 6.2.

2D-PEEC Model For analyzing the AC internal impedance of the conductor with the 2D-PEEC method, the cross section is subdivided according to Figure 4.11 and solved

Chapter 6 Simulation Results and Measurements

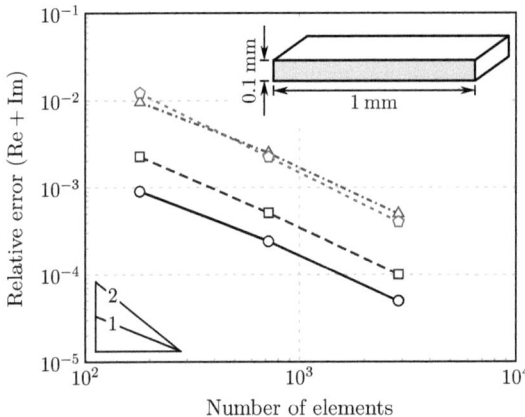

Figure 6.7: 2D convergence analysis of the conductor discretization for different frequencies. The reference simulation is performed with 11 520 current cells. (s. also footnote 1)

via (6.1) and (6.2a) for an initial mesh setting. In order to quantify the discretization error, each basic current cell is subdivided into four equally spaced rectangles and the system is solved again with the refined mesh. This procedure is repeated several times up to a finest resolution which is used as a reference solution for the convergence study.[1] The results are presented in Figure 6.7 for four different frequencies. All curves show a convergence order of approximately one whereas the different offsets are determined by the initial discretization. If this discretization is fine at the boundary region and coarse at the interior, the discretization error is almost equally distributed for all elements at high frequencies. On the other hand, at low and medium frequencies the error is dominated by the coarse elements in the interior of the conductor. For an equidistant initial discretization, the situation is contrary.

In Figure 6.8a, the normalized resistance with the finest mesh setting presented in Figure 6.7 (2 880 current cells) is visualized over a broad frequency range. According to the circular cross section, the resistance increases above a certain frequency limit from which the influence of the eddy currents becomes significant. In contrast to the circular cross section, an intermediate region in between 100 kHz and a few MHz can be observed. In this region, the increase on the resistance is lower than the high frequency limit which is well known to be proportional to $\sqrt{\omega}$. This property can be explained by the cross section of the conductor which has a ratio of the width by the thickness of 10 in this example. In the intermediate frequency range, the skin effect is distinct only in the width direction while the current density is almost constant in the thickness direction.

---
[1] Only three points are visualized in Figure 6.7 due to the following reasons. First, memory limitations of the actual code implementation prohibit a further refinement of the chosen reference solution. Second, too coarse mesh settings cause very large errors of little relevance. Third, intermediate values would demand additional implementation effort since an interpolation of subdivided rectangles would be necessitated.

6.1 Cylindrical Conductor

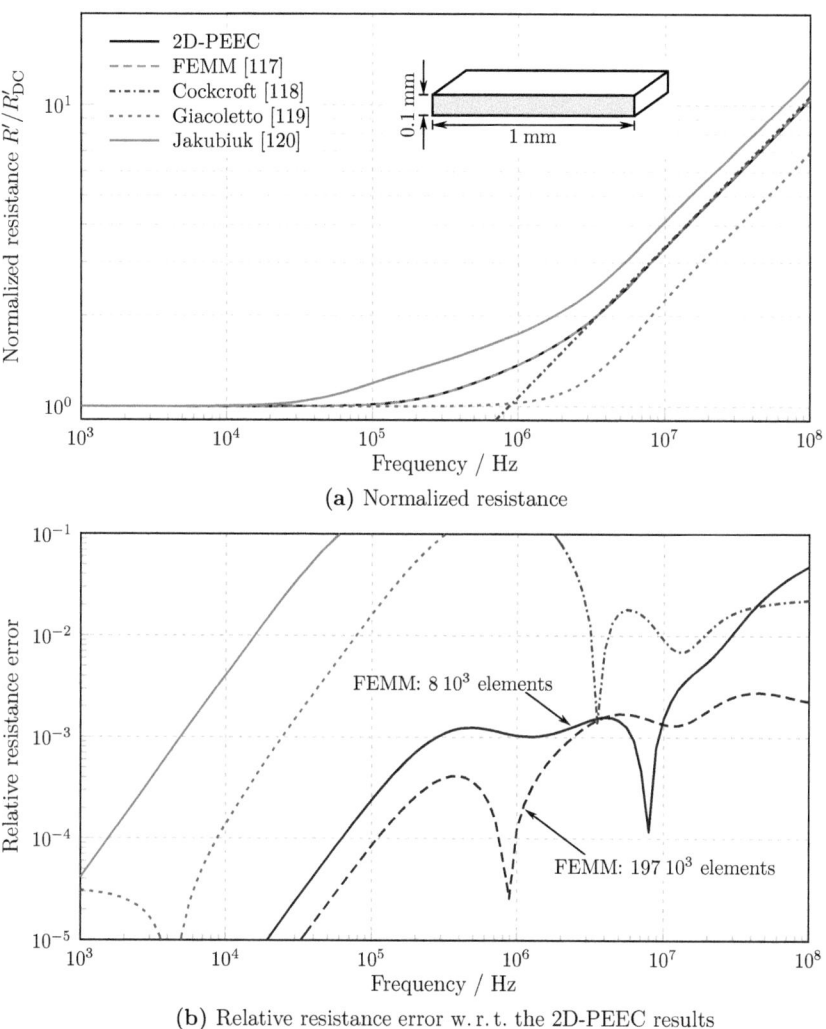

(a) Normalized resistance

(b) Relative resistance error w.r.t. the 2D-PEEC results

Figure 6.8: Frequency dependent resistance of the rectangular conductor computed with multiple approaches. The curves of the 2D-PEEC and the FEMM approaches agree in almost the whole considered spectrum with an error not exceeding 0.3 % if a high resolution of the FEMM mesh is chosen. The errors of the approximation techniques can only be used as rough estimations since the errors exceed 10 % for wide frequency ranges. For high frequencies, only the Cockcroft approach seems to be suitable.

Chapter 6 Simulation Results and Measurements

Comparison results performed with a conductor of square cross section confirm this explanation since in the square case, the intermediate range is not observable. In order to corroborate the 2D-PEEC results, the same setup is simulated with the 2D-FEM solver FEMM [117]. Errors for two different discretization settings are visualized in Figure 6.8b.

Attempts to Analytical Solutions Although the 2D-PEEC and the 2D-FEM results show an excellent agreement, in the following, an overview of different analytical attempts for calculating the frequency-dependent behavior of a single conductor with rectangular cross section is presented. For this reason, the computed analytical results are compared with the numerical results from the previous considerations for the same test conductor.

For at least 100 years, researchers have been investigating on the high frequency current distribution of a single infinitely long conductor with a rectangular cross section. The dimensions are equivalent to Figure 6.6 while the excitation is given in terms of the total current $\underline{I}_0$. In contrast to the circular cross section in which the current distribution is a function of the ϱ-coordinate only, in the rectangular case, the current density $\underline{J}_z(x,y)$ is a function of the two variables x and y. This results in an analytical solution which is much more complex compared to the circular case. The common approach for solving problems with a dependence of multiple coordinates is to apply a separation of variables in the form of a series representation. The unknown coefficients are then determined by applying of some beneficial boundary conditions. However, for this special kind of problem it is very difficult to determine the conditions at the boundaries of the conductor since neither the potential nor the field components are known a priori.

Giacoletto Model: Some contributions overcome this difficulty by fixing the fields to a constant value at the conductor edges. By doing so, an approximated closed-form solution is obtained which has been proposed by Giacoletto [119, eq. (63)]. Although this solution might have been presented by earlier authors, it is referred to as *Giacoletto* approach in the following. As can be seen from Figure 6.8a, the resistance is underestimated when using this method.

Cockcroft Model: In order to avoid the difficulty of finding the boundary condition at the conductors edges, conformal mapping techniques can be used to transform the rectangular boundaries to a unit circle by means of a Schwarz-Christoffel mapping. This is advantageous since the available analytical solution of the circular cross section can be transformed to the solution of the rectangular domain. This approach has been applied by Cockcroft in 1929 [118]. In there, a solution to the frequency-dependent resistance for the high-frequency limit is presented in which the current is assumed to flow on the surface of the conductor only. Thus, an equivalence with the surface charge of an electrostatic problem can be shown. By multiplying the equivalent surface current density with the skin depth as an effective penetration, the frequency-dependent resistance equation can be derived. This technique is named *Cockcroft* model in the following. As can be verified by Figure 6.8, the Cockcroft approach provides the best approximation results for high frequencies. It should be mentioned that the conformal transformation results in a singularity of the surface current density at the corners of the conductor which does not seem to reflect the physical volume current density in a correct manner. This could explain the relatively constant errors at high frequencies in Figure 6.8b.

Jakubiuk Model: In 1976, Jakubiuk and Zimny [120] presented a full spectrum method which is also based on the Schwarz-Christoffel mapping. The basic idea is to excite the conductor with a rectangular current surge at an arbitrarily chosen initial moment. The immediate rise of the total current is assumed to correspond to a frequency approaching infinity. This in turn implies a skin depth of zero and a surface current at the initial moment only. The surface current density can be computed according to the above Cockcroft model. This surface current is used as the initial condition for the series representation of the total volume current distribution. Afterwards, the convolution technique is applied to transfer the known current distribution of the unit step excitation to the current distribution of an harmonic excitation. An implementation of this technique shows a slow convergence of the resulting double series representation [120, eq. (33)] with a cumbersome coefficient evaluation. Furthermore, the resistance of (2.39) must be evaluated numerically from the obtained current distribution which even more complicates the overall evaluation. Numerical results in Figure 6.8 show an overestimation of the resistance in the Jakubiuk model.

Groß Model: Another attempt to solve for the current distribution inside the cylindrical conductor with a rectangular cross section has been presented by Groß [121] in 1940. Groß proposes an iterative computation of the integral formulations of the Faraday's law and Ampere's law which bypasses the difficulty of finding the boundary conditions at the conductor edges. Although a theoretical computation scheme for the current distribution is demonstrated in [121], no results to the iterative process are presented. Due to the iterative character of the method as well as the complexity of the equations in combination with a numerical evaluation of the resistance, results of this approach are not appended to Figure 6.8.

Summing up, the analytical expressions can be used as rough estimations only whereas the Cockcroft model [118] seems to be most suitable for the high frequency limit while being computationally not expensive.

6.1.4 Rectangular Cross Section and Finite Length

In contrast to the previous results in which only 2D problems have been regarded, in this section the length of the conductor is chosen as finite. This equals a transition from the 2D-PEEC method to the MQS-PEEC method. Analytical or numerical results obtained

Chapter 6 Simulation Results and Measurements

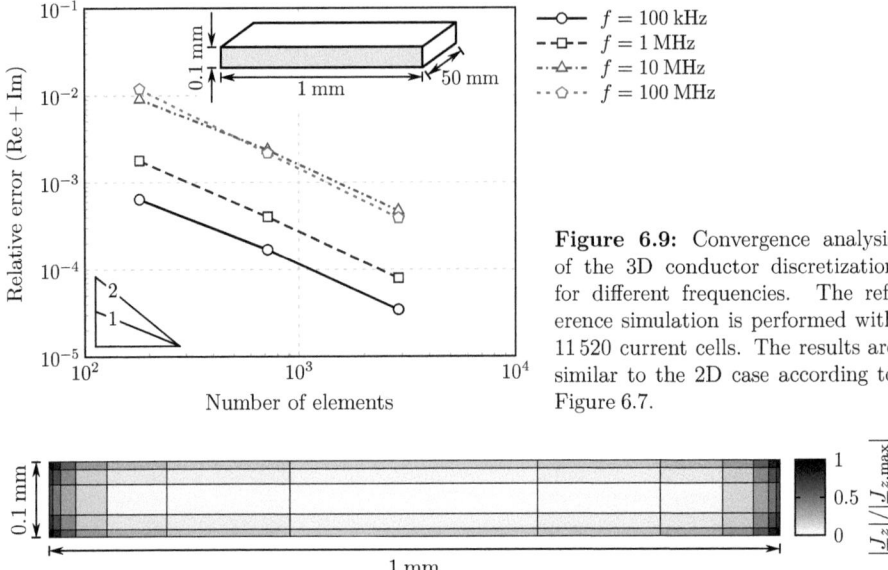

Figure 6.9: Convergence analysis of the 3D conductor discretization for different frequencies. The reference simulation is performed with 11 520 current cells. The results are similar to the 2D case according to Figure 6.7.

Figure 6.10: Exemplary current distribution at 10 MHz of the discretized conductor of Table 6.2 with $l = 50\,\text{mm}$, $\chi = 2$ (skin factor), $\delta_{wt} \leq \delta/2$ (mean width of the corner elements), $N_w = 13$ and $N_t = 5$ elements. Thus, the number of current elements is $N = 65$.

by other 3D solvers are not presented in this section. Instead, the MQS-PEEC solution is compared with the 2D-PEEC approach as being a limiting case for long conductors. A comparison with a 3D-FEM solver is focused on in sections 6.2 and 6.3.

PEEC Model According to the above, a convergence analysis is performed in Figure 6.9 for a test conductor with the cross sectional dimensions as before (cf. Table 6.2). In addition, the conductor length of $l = 50\,\text{mm}$ is chosen which is a realistic choice for PSCs in IPT systems. The chosen length emphasizes the high aspect ratios of the cells that might occur in PEEC systems. The results are similar to the 2D case according to Figure 6.7 which can be explained by the fact that only the mutual element couplings of (4.12a) have been substituted by (4.29). The meshing and solving algorithms remain unchanged. An exemplary current distribution obtained by the MQS-PEEC method is presented in Figure 6.10. In order to compare the results from the MQS-PEEC approach with the 2D-PEEC counterpart, the length of the conductor is increased. For each length, the per-unit-length impedance is approximated in the 3D case by dividing the computed value by the length. In Figure 6.11, the deviation

6.1 Cylindrical Conductor

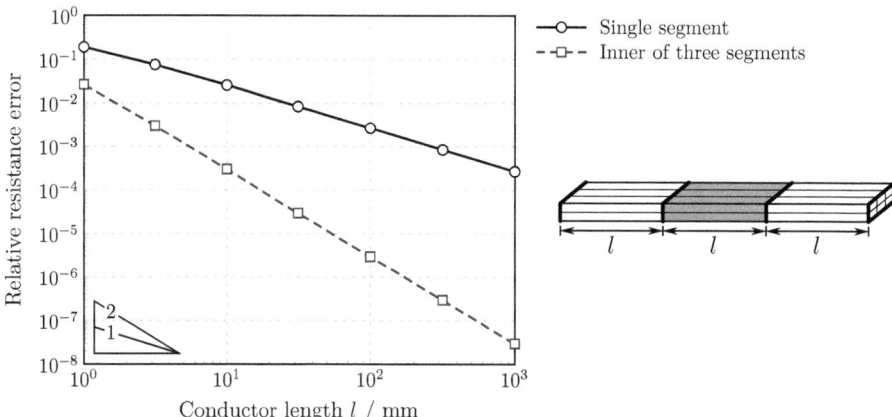

Figure 6.11: Comparison of the test conductor of Table 6.2 between 2D-PEEC and MQS-PEEC normalized to the length l at $f = 10\,\text{MHz}$. As expected, the error decreases when the length is increased. The error converges faster for a series connection of three segments when only the resistance of the inner segment is computed via a discrete form of (2.39). The discretization settings of this example are chosen according to Figure 6.10.

from the 2D limit is visualized for different lengths. Obviously, the influence of the edge effects decreases when the conductor is enlarged. This effect is even more accelerated if three conductors are connected in series while only the per-unit-length resistance of the inner segment is computed via a discrete form of (2.39). This is due to the fact that the field distortion at the start and end points of the conductor are almost non-existent for the inner segment.

At this point, it should be pointed out that the subdivision of the rectangular cross section via the skin factor χ as defined in Figure 4.11 is an important meshing factor. This is because too large values increase the discretization error while too small values unnecessarily increase the simulation time. Case studies are presented in appendix C on page 177. In there, it is shown that a value of χ in between two and three yields acceptable results within an error of approximately 1 %.

6.1.5 Sensitivity Analysis

In this section, the derivatives of the conductor impedance w. r. t. the design parameters p_n are focused on. In the single conductor example with a rectangular cross section, the design parameters p_n may include any of the geometrical parameters w, t and l as well as the frequency f and the conductivity κ. According to (6.2b), the derivatives of the inductance matrix $\partial \mathbf{L}/\partial p_n$ of (5.8) must be set up. It is assumed that the branch currents $\mathbf{i}_b$ of the initial system and the derivatives of the partial resistance matrix $\partial \mathbf{R}/\partial p_n$ with the definitions of (4.11) have already been computed. From all mentioned design parameters, the most challenged derivative is $\partial \mathbf{L}/\partial w$ since generally all mesh cells are shifted and scaled when the width is perturbed. This results in a dense matrix of derivatives.[2] For this reason, the following considerations focus on the derivatives w.r.t. the width, i.e. $\partial \underline{Z}_0/\partial w$. More practical sensitivities of a single coil will be discussed in section 6.2.3 on page 142.

Convergence Study via Exact Sensitivities In the first study, the exact sensitivities with the expressions according to appendix A.2 on page 162 are computed for the three different approaches visualized in Figure 5.2. A convergence analysis is shown in Figure 6.12 for the exemplary setup according to Figure 6.9. In Figure 6.12a, the elements are uniformly perturbed which is the most general and most expensive approach at the same time. This is due to the fact that for the matrix fill of $\partial \mathbf{L}/\partial w$, all entries differ from zero since all elements are shifted and scaled. Again, the vertical shift of the four different frequency curves is caused by the initial discretization.

In contrast to the uniform perturbation, the Inner-Layer Concept (ILC) presented in Figure 6.12b demands only about half of the element computations because the right and left blocks of elements are shifted as a group. This results in $\partial L_{mn}/\partial w = 0$ for these entries and consequently no computational costs. The obtained errors of the ILC are comparable to the uniform perturbation which makes this method well suited for skin-effect applications.

In Figure 6.12c, the results of the Boundary-Layer Concept (BLC) are presented. When comparing the results with the two methods from above, the obtained errors are significantly larger, especially at high frequencies where the current is concentrated at the surface. Thus, the applicability of this method to skin-effect problems is limited.

Finite Difference Approximations For practical applications, the implementation effort to express the exact derivatives for all possible configurations may be difficult if not impossible to determine because many different cases must be considered.[3] In order to overcome this difficulty, Finite Difference (FD) approximations can be applied which provide an approximation of the exact derivatives. This is of advantage since almost no additional code has to be implemented. On the other hand, a numerical step size Δp_n needs to be defined which introduces additional error components. Thus, the derivatives

---

[2] The computation of $\partial \mathbf{L}/\partial t$ is equivalent because the width and the thickness of the conductor can be interchanged.
[3] A possible way is to use the method of Automatic Differentiation (AD), e.g. [113].

6.1 Cylindrical Conductor

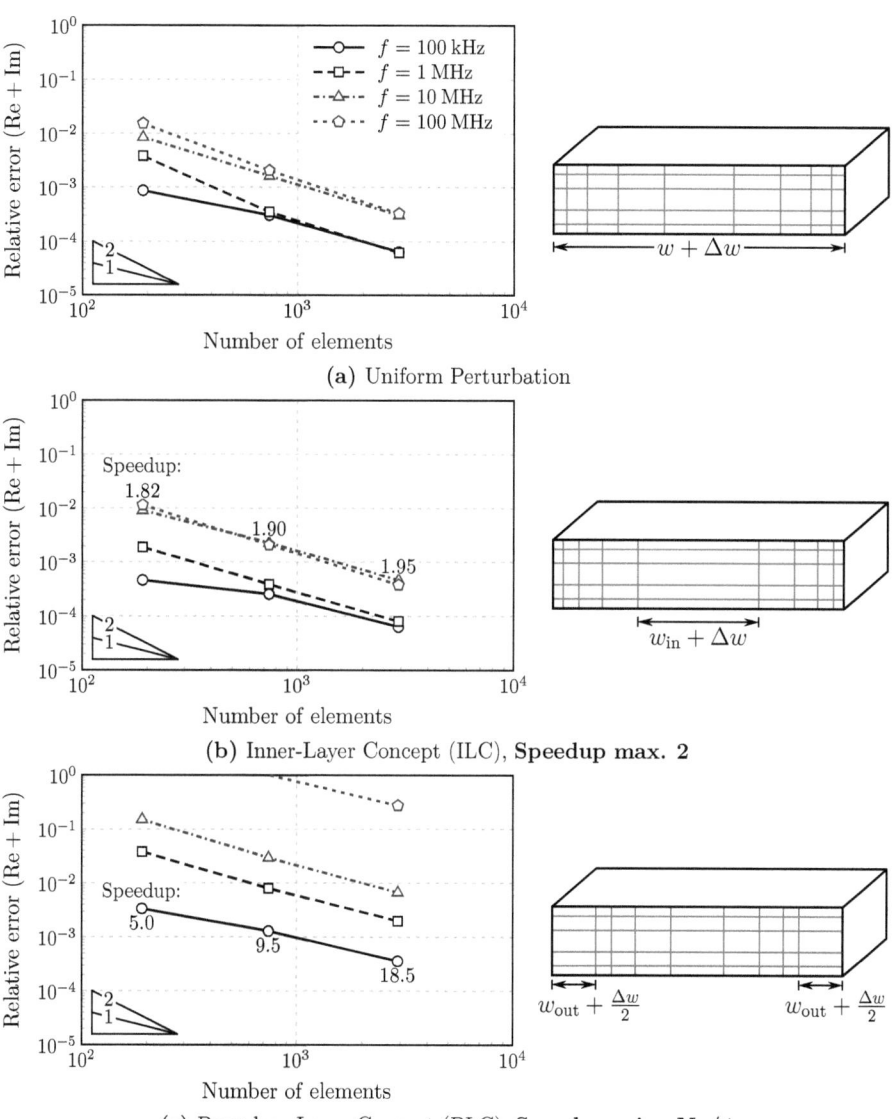

Figure 6.12: Convergence analysis of $\partial \underline{Z}_0/\partial w$ for three different methods. While the errors of the upper two approaches are comparable, the BLC is not applicable especially for high frequencies. The dimensions and mesh settings are chosen according to Figure 6.9.

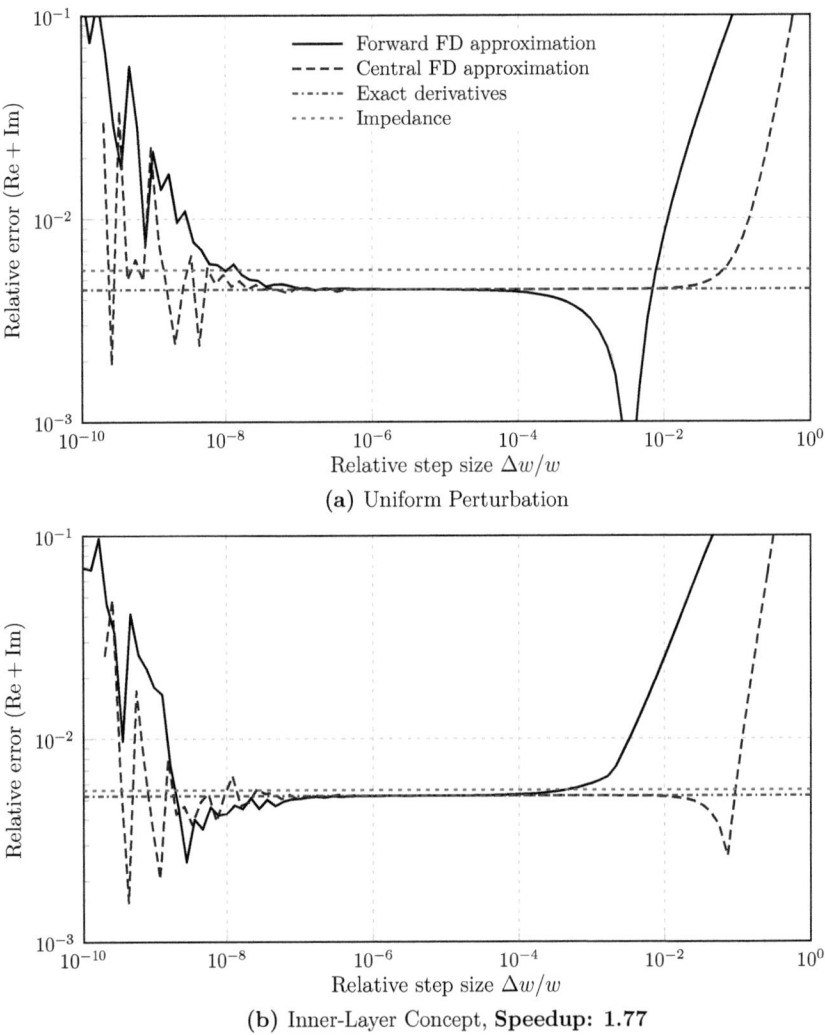

Figure 6.13: Finite difference approximations for the extraction of $\partial \underline{Z}_0 / \partial w$ via the adjoint sensitivity analysis compared with exact derivatives obtained by appendix A.2. The test setup and the discretization settings are equivalent to Figure 6.10. Compared to the central FD approximation, the forward FD approximation requires a step size of about two orders of magnitude smaller for obtaining similar results. If the step size is chosen too small, numerical noise occurs and accurate results are no longer maintained.

of the partial inductances can be approximated as

$$\frac{\partial L_{mn}}{\partial p_n} \approx \frac{L_{mn}(p_n + \Delta p_n) - L_{mn}(p_n - \Delta p_n)}{2\Delta p_n}, \quad \text{(Central FD approximation)} \quad (6.4\text{a})$$

$$\frac{\partial L_{mn}}{\partial p_n} \approx \frac{L_{mn}(p_n + \Delta p_n) - L_{mn}(p_n)}{\Delta p_n}. \quad \text{(Forward FD approximation)} \quad (6.4\text{b})$$

In Figure 6.13, the FD approximations (6.4) are computed for the width w. The resulting derivatives are substituted in (6.2b) in order to obtain an approximation of $\partial \underline{Z}_0/\partial w$ for the test conductor as specified in Figure 6.10. The relative step size $\Delta w/w$ is varied in a wide parameter range. In the figure, the error w.r.t. the reference values obtained by the convergence study according to Figure 6.12 is shown. In addition, the constant values of the exact derivatives are visualized from which follows that relative step sizes of in between 10^{-4} to 10^{-2} can be used to obtain acceptable errors.

Obviously, the central FD approximation is more accurate as it converges faster to the exact value. On the other hand, the computation time is as twice as much as the forward FD approximation caused by an additional matrix setup. When comparing the effort of the central FD approximation with the exact sensitivities, both approaches are comparable. This is due to the fact that two additional matrix fills at $w + \Delta w$ and $w - \Delta w$ are required in contrast to a single evaluation of $\partial \mathbf{L}/\partial w$ in (A.6) which is approximately twice as costly as setting up $\mathbf{L}$.

6.2 Printed Spiral Coil

The second section of the results chapter leads to an optimized design of a rectangular multi-turn PSC which is often used in IPT systems. First, a conductor bend with two conductors in a right angle to each other is analyzed under stationary conditions. This allows for optimizing the 2D mesh settings of the PEEC models since an analytical reference solution is available for this case. Next, a single-turn coil is considered with the MQS-PEEC method. Also, a comparison w.r.t. full-wave FEM results is performed in terms of accuracy and effort. In the last part of this section, a multi-turn PSC is designed and optimized with the LQS-PEEC method whereas the final design is compared with a FEM reference simulation as well as with measurements. Additionally, reduced network models are provided and a sensitivity analysis with the fabrication tolerances obtained by the manufacturer is performed. The finalized coil design will be used in section 6.3 as part of an inductively coupled RFID antenna system.

Chapter 6 Simulation Results and Measurements

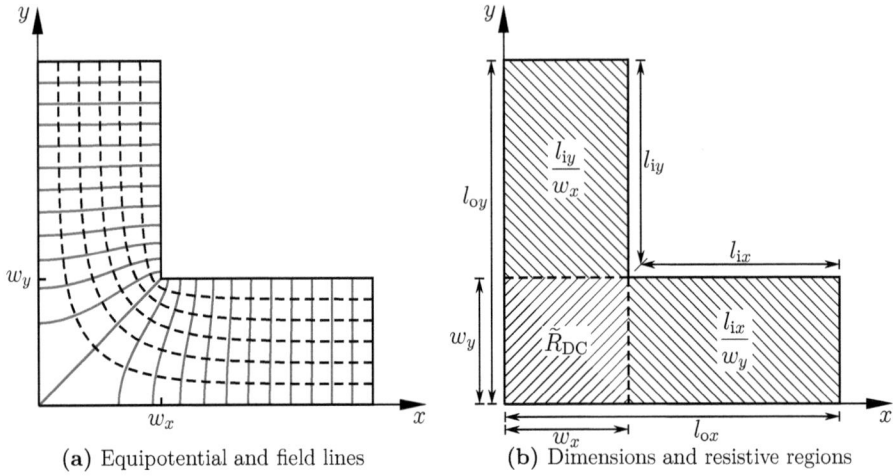

(a) Equipotential and field lines   **(b)** Dimensions and resistive regions

Figure 6.14: (a) Electrostatic field distribution obtained by the conformal-mapping technique. (b) Resistive regions in a rectangular conductor bend allowing the computation of the resistance as $R_{\text{DC}} = 1/(\kappa t)(l_{\text{i}x}/w_y + l_{\text{i}y}/w_x + \tilde{R}_{\text{DC}})$ with the correction term $\tilde{R}_{\text{DC}}$ defined in (6.5).

6.2.1 Two Conductors Connected in Right Angle

As stated before, the analysis is started with a single rectangular conductor bend in order to verify the 2D mesh settings from section 4.5.2. Since an analytical solution is available for the DC case, the DC-PEEC method of section 4.4.3 is applied.

Analytical DC-Resistance by Conformal Mapping The DC resistance of a rectangular conductor bend can be found by means of the Schwarz-Christoffel transformation whereas it is assumed that the unconnected ends of the conductors extend towards infinity. In Figure 6.14a, the mesh transformed by conformal mapping is presented. The solid lines represent constant potential values while the dashed lines visualize the electric field lines. It is shown in appendix B on page 169 that the DC resistance R_{DC} of the considered corner element can be calculated by taking the resistances of the inner dimensions as in Figure 6.14b and adding a correction term $\tilde{R}_{\text{DC}}$, defined as

$$\tilde{R}_{\text{DC}} = \frac{w_y}{w_x} + \frac{2}{\pi}\ln\left(\frac{w_x^2 + w_y^2}{4\,w_x w_y}\right) + \frac{2}{\pi}\frac{w_x^2 - w_y^2}{w_x w_y}\arctan\left(\frac{w_y}{w_x}\right), \quad (6.5\text{a})$$

6.2 Printed Spiral Coil

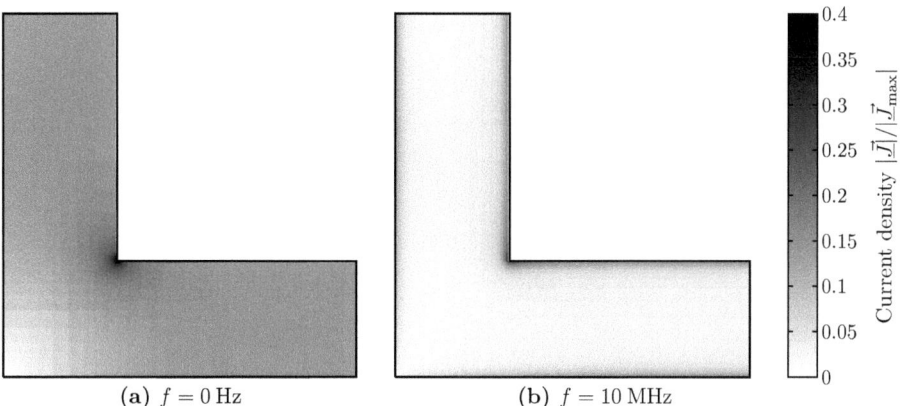

(a) $f = 0\,\text{Hz}$ (b) $f = 10\,\text{MHz}$

Figure 6.15: Simulated current distribution with (a) the DC-PEEC method and (b) the MQS-PEEC method for the rectangular conductor bend with the dimensions according to Table 6.3. A singularity occurs at the innermost edge of the corner in both cases. At high frequencies, the skin effect additionally influences the current distribution.

with the widths w_x and w_y according to Figure 6.14b. If both conductors hold the same width $w_x = w_y = w$, the correction term simplifies to

$$\tilde{R}_{\text{DC}} = 1 - \frac{2\ln 2}{\pi} \approx 0.5587, \qquad \text{for} \qquad w_x = w_y, \tag{6.5b}$$

which has already been stated in [122]. Due to the geometrical assumptions, this term is exact only when the lengths l_{ix} and l_{iy} displayed in Figure 6.14b approach infinity. However, the relative error decreases exponentially for an increasing length-by-width ratio. In Figure B.4 of appendix B it is shown that the error is numerically negligible if the ratio of the length by the width is larger or equal to 10. Even for a ratio of two, the error is typically below 0.1 %. This allows for computing the DC resistance of a multi-turn PSC with the geometrical dimensions according to Figure 3.5 as

$$R_{\text{DC}} = \frac{1}{\kappa t}\left(\frac{l_i}{w} + N_c\, \tilde{R}_{\text{DC}}\right), \tag{6.6}$$

where κ is the conductivity, t the thickness, w the width and l_i the accumulated inner length of the traces of the PSC. The number of corners is denoted by N_c.

DC-PEEC Model The analytical DC resistance of (6.6) is compared with the DC-PEEC approach of section 4.4.3 for a single, symmetrical conductor bend with the parameter values of Table 6.3. The dimensions can be verified by Figure 6.14b. The

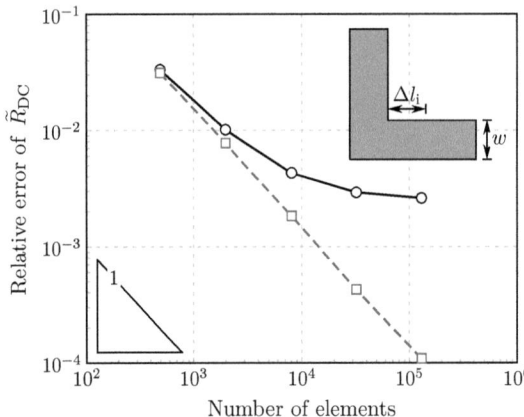

Figure 6.16: Convergence analysis of the DC resistance of the corner element. If the 2D mesh extension Δl_i according to Figure 4.8 is chosen approximately to the conductor width, the error remains above the threshold of approximately 0.2 %. An extension of the 2D discretization of $\Delta l_i \approx 2\,w$ is sufficient in this example.

meshing of the rectangular bend is done according to Figure 4.8 whereas the 2D extension Δl_i is varied. In Figure 6.15, the current distribution is visualized for two different frequencies. For the DC case, the current density is almost constant inside the long conductors whereas it concentrates at the innermost corner of the 90° bend. At high frequencies, an additional current densification towards the boundaries can be observed due to the skin effect.

Name	Value
l_o	5 mm
w	0.8 mm
t	35 µm
κ	58 10^6 S/m

Table 6.3: Parameters of the test setup

The DC resistance of the 2D-PEEC model is compared with the analytical reference solution (6.6) in terms of a convergence analysis. In each refinement level, the bars are subdivided by four elements each. Since the numerical error depends on the lengths of the straight connected conductors, the resistance parts belonging to the long conductors are subtracted from the overall resistance. This allows for focusing on the correction term $\tilde{R}_{\mathrm{DC}}$. In Figure 6.16, results are plotted for two different extensions of the 2D discretization according to Figure 4.8. It can be seen that it is sufficient to enhance the 2D discretization Δl_i up to two times the width in order to obtain relative errors below 10^{-4}. When comparing the number of elements presented in Figure 6.16 with the number of elements from previous convergence analyses, it is obvious that much more cells can be used in the DC-PEEC case. This is due to the fact that no dense matrix of partial inductances needs to be set up, thus resulting in a sparse system (4.26).

As a result, the proposed meshing algorithm presented in Figure 4.8 is capable of modeling the 2D-current distribution in the DC case in a correct manner. For obtaining a sufficient accuracy, the 2D extension towards the long conductors can be chosen in the region of in between $w \leq \Delta l_i \leq 2w$. Results for the AC case will be presented in the next section for a single-turn PSC.

6.2.2 Rectangular Single-Turn Coil

In this section, the MQS-PEEC approach is applied to a single-turn coil. The results are compared with a commercial FEM solver since exact analytical expressions are not available for this kind of application. The single-turn coil acts as a preliminary stage of the multi-turn coil in which also the LQS-PEEC method with capacitive effects will be considered. The parameters of the single-turn coil are presented in Table 6.4 whereas the geometry can be viewed in Figure 6.17. The following simulations will be performed at the frequency of $f = 10\,\text{MHz}$.

Name	Value
l_x	10 mm
l_y	8 mm
w	0.8 mm
t	35 µm
s_port	30 µm
κ	$58\,10^6$ S/m

Table 6.4: Test coil

CST Microwave Studio Solver Settings In order to compare the MQS-PEEC results with a reference solution, the frequency domain full-wave FEM solver of the CST MICROWAVE STUDIO® suite [29] is applied. This solver is preferred over the time domain solver due to the estimated SRF of the analyzed coils at a relatively high quality factor. The FEM simulation of IPT antenna systems is challenged due to the occurring high ratios of different cell sizes ranging from micrometers for cells near the conductor surfaces up to centimeters for cells located distant from the coil. The geometry of the single-turn coil as well as an excerpt of the tetrahedral mesh are visualized in Figure 6.17. The port is modeled as a discrete current port which is connected via two Perfect Electric Conductor (PEC) bricks to the coil. The chosen port separation of $s_\text{port} = 30\,\text{µm}$ is small compared to the other dimensions of the coil in order to reduce the port influence on the solution. The background material is modeled as free space whereas the computational domain is terminated by an electric boundary condition. In order to save computational effort, two symmetry planes are applied, thus reducing the overall volume to a quarter.

In order to accurately model the skin-effect losses inside the conductors at medium frequencies, a fine volume discretization is required. For minimizing the discretization effort, the conductors are assembled as different sized bricks with a local mesh setting according to their position. This allows a fine mesh near the surface of the conductors while the interior can be discretized with larger elements (cf. Figure 6.17). A global mesh setting enables a smooth transition from the fine mesh to regions with coarser discretization. This feature is especially used for the discretization of the background material which is not shown in Figure 6.17 for clarity reasons. The FEM results are

Chapter 6 Simulation Results and Measurements

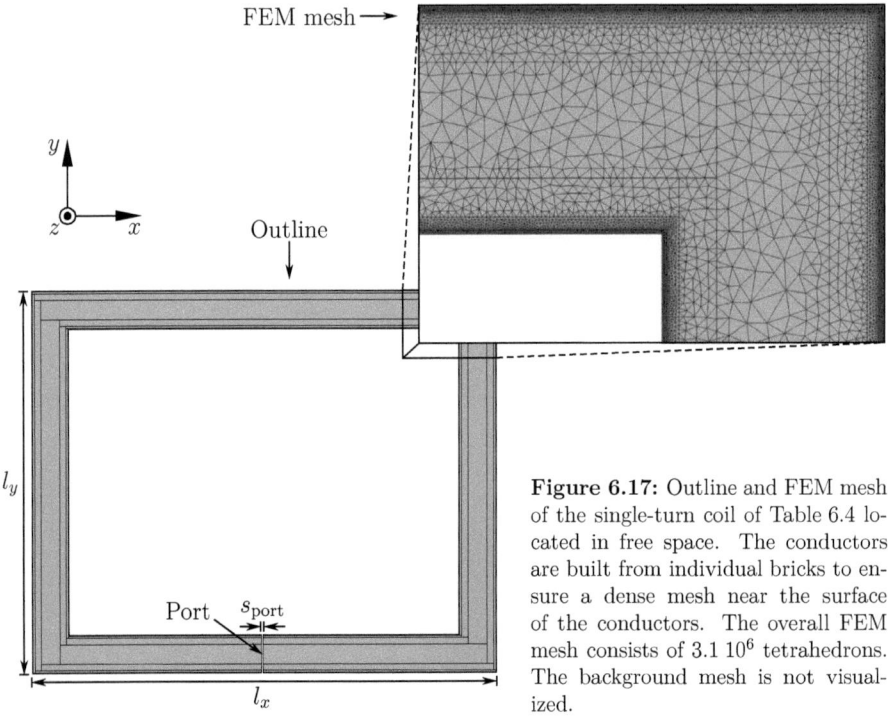

Figure 6.17: Outline and FEM mesh of the single-turn coil of Table 6.4 located in free space. The conductors are built from individual bricks to ensure a dense mesh near the surface of the conductors. The overall FEM mesh consists of $3.1 \cdot 10^6$ tetrahedrons. The background mesh is not visualized.

carried out with an iterative solver[4] for three different orders of basis functions. A global refinement parameter is set up which allows for determining the mesh density of the simulation. The parameter also minimizes the influence of the electric boundary on the results. Both properties are obtained by simultaneously influencing the conductor discretization on the one hand and the size of the bounding box on the other hand. For each order of basis functions, the refinement factor is increased until either the memory limit of the workstation or the maximum acceptable simulation time (about one week) is reached.

The results are presented in Figure 6.18 for the frequency of 10 MHz. In detail, the deviation is plotted w.r.t. the finest MQS-PEEC solution with the solver settings explained in the next paragraph. As expected, the error decreases with an increasing order of basis functions. Especially the resistance error is remarkably low which is due to the good approximation of the high current distribution at the surface of the conductors by means of higher order basis functions. In contrast to the resistance error,

---
[4]The solver accuracy is set to 10^{-8}.

6.2 Printed Spiral Coil

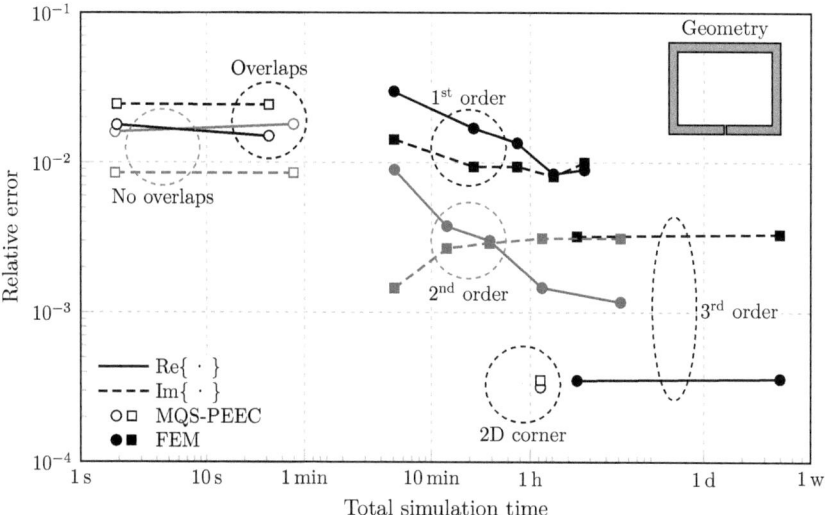

Figure 6.18: Comparison of full-wave FEM and MQS-PEEC results at 10 MHz as a function of the overall simulation time. The MQS-PEEC results with the 2D-corner discretization and a refinement factor of two are chosen as the reference values. While the PEEC results with the simplified corner discretization (s. Figure 4.7) can be obtained in seconds, their error is not significantly reduced when refining the mesh inside the conductors. The most accurate FEM results are obtained by using third order basis functions as they best approximate the high current density near the surface of the conductors. The imaginary parts of the FEM results show a systematic deviation from the MQS-PEEC reference solution. This can be explained by the magnetic field created by the discrete port which is neglected at the PEEC results. A further explanation is the non-vanishing electric energy of the full-wave FEM results which influences the reactive component of the port impedance. Even though the implementations of the different FEM and PEEC codes cannot be compared directly, a significant speedup can be obtained when using the PEEC method and accepting relative errors of a few percent.

the deviation of the inductance obtained by the full-wave FEM solution and the MQS-PEEC reference simulation converges to approximately 0.3 % which can be explained as follows. The magnetic energy[5] of the infinitesimal thin discrete port current is accounted for in the FEM results whereas it is not modeled in the MQS-PEEC approach. Moreover, the neglect of the displacement current in the MQS-PEEC approach which corresponds to zero electric energy differs from the full-wave solution in which the part of the electric energy is incorporated in the inductance.[6]

---

[5]The magnetic energy is directly related to the inductance as can be seen from (2.41) for instance.
[6]This can be verified by (2.58) for example.

Chapter 6 Simulation Results and Measurements

MQS-PEEC Solver Settings Prior to comparing the FEM results with the results obtained by the PEEC method, a few additional hints about the solver settings are presented. For the MQS-PEEC model, the coil as visualized in Figure 6.17 is partitioned into five straight conductor segments while the cross section of each segment is subdivided into $N_w = 11$ and $N_t = 3$ elements. The subdivision is performed according to (4.36) in which a skin factor of $\chi = 2$ is used. This ensures a sufficient discretization of the outermost segment having $\delta_w = 8.51$ µm and $\delta_t = 8.75$ µm (s. Figure 4.11) at a skin depth (3.10) of $\delta = 20.9$ µm at the frequency of 10 MHz. The four corners of the coil are discretized in three different ways as visualized in Figure 4.7. Thus, the numbers of elementary current cells are 6 333 for the fully discretized corner ($\Delta l_i = 2w$) and 165 for the both 1D approximations. In order to analyze the convergence of the results, a second discretization is set up with each element constituting only half of the size which results in $N_w = 22$ and $N_t = 6$ subdivisions. This leads to a number of unknown currents of 51 732 in the 2D case and 660 in the 1D cases, respectively. After computing the matrices of partial resistances and inductances as well as the nodal connectivity matrix, the system (4.25) is solved for the branch currents and node potentials. Applying (4.21b) allows for computing the port impedance of the coil.

The results of the three different corner discretization approaches displayed in Figure 4.7 are appended to Figure 6.18. The 2D corner discretization is a single pair of values only since the fine mesh is used as the reference value. While the error of the corner discretization is well below 0.1 %, the errors of the 1D corner approximations amount a few percent. It should be mentioned that the error of both simplified versions does not decrease if a finer cross sectional mesh is used. This can be explained by the fact that the error is dominated by the simplified discretization in the region of the corner.

In [123], a more detailed comparison of the three approaches shows that the results obtained by the non-overlapping approach Figure 4.7c are more accurate compared to the overlapping approach Figure 4.7b especially for multi-turn coils. For this reason, only the non-overlapping discretization routine will be used in the following. It should be noted that the use of the simplified 1D-corner discretization is justified in most cases. This is due to the fact that the practicable cross-section discretization method of appendix C on page 177 typically produces errors in the range of 1 %. For this reason it is not appropriate to discretize the corners with a 2D mesh leading to a much higher precision than 1 %.

Comparison of the FEM and the PEEC Results The simulations presented in Figure 6.18 are performed on a computer with a 64 bit architecture, 64 GB RAM and a 3.0 GHz quad-core processor. The FEM and MQS-PEEC results are compared in terms of the overall simulation time which is the only feasible approach. This is due to the facts that not only two different numerical methods are compared but also the

6.2 Printed Spiral Coil

Name	Value	Tolerances	
l_x	50 mm	—	—
l_y	60 mm	—	—
N_{turn}	5	—	—
w	1.636 mm	$-1.8\,\%$[a]	—
s	0.778 mm	—	—
t	35 µm	$-15\,\%$	$+20\,\%$
h	1.55 mm	$-7.1\,\%$	$+13.6\,\%$
κ	$58\,10^6$ S/m	$-10\,\%$	$+10\,\%$
ε_r	4.6	$-10\,\%$	$+10\,\%$
$\tan\delta$	0.018	$-10\,\%$	$+10\,\%$

[a]equals -0.03 mm at the nominal width

Table 6.5: Test setup of the multi-turn PSC in a standard PCB technology. The gray shaded values are determined during the optimization process. The meaning of the geometrical dimensions is visualized in Figure 3.5 while the actual geometry is plotted in Figure 6.20a. The material properties and their tolerances are obtained from datasheets provided by the PCB manufacturer.

implementations of the codes differ. Nevertheless, a significant speedup of the MQS-PEEC results is gained under the premise that errors of a few percent are acceptable. If a high precision is required, the PEEC method with a 2D corner discretization or the FEM with third order basis functions can be set up.

A further result of Figure 6.18 is the fact that the MQS approximation is sufficient for this setup. This is confirmed by (2.26) since the geometrical dimensions are sufficiently below the border of quasi-stationary assumptions which is about 4.8 m in this case.

6.2.3 Rectangular Multi-Turn RFID Antenna

In this section, a multi-turn RFID antenna is designed and optimized for a given transponder IC which is characterized by its input impedance. The system is operated at the frequency of 13.56 MHz and the antenna is mounted on a standard PCB. Throughout the design process, the PEEC method is used as the simulation tool. In particular, the MQS-PEEC approach is now extended by the LQS-PEEC approach. This is necessary for two reasons. First, the capacitive cross couplings of the wires have a reasonable impact on the system behavior. Second, the joint simulation of both LQS-PEEC and MQS-PEEC methods is used to extract the reduced circuit models of section 3.3.2. These circuit models are admirable for the system design based on the transformer concept. The results obtained by the LQS-PEEC method will be verified by FEM simulations and measurements carried out on an impedance analyzer. At the end of this section, a sensitivity analysis will be performed with the design parameter tolerances obtained by the PCB manufacturer.

In Table 6.5, the shape and technology parameters values of the 5-turn PSC are presented. The dimensions are chosen according to Figure 3.5 while the optimized geometry is visualized in Figure 6.20.

Chapter 6 Simulation Results and Measurements

Optimization of the Antenna Impedance In the following it is aimed to design and optimize the rectangular transponder antenna in standard PCB technology. The outer dimensions of the PSC are pre-specified with $l_x = 50\,\text{mm}$ and $l_y = 60\,\text{mm}$. Moreover, the transponder IC input impedance is characterized by the load impedance consisting of C_{Load} parallel to R_{Load} with the values of Table 6.6. The equivalent circuit of the setup is chosen according to Figure 3.12 whereas a discrete matching network capacitance should be avoided, i.e. $C_{\text{MN}} = 0$. This demands that the imaginary part is matched to the load impedance at the working frequency of $f_0 = 13.56\,\text{MHz}$.[7] A convenient way is to convert the load impedance at the working frequency to an equivalent series connection

$$Z_{\text{Load}}(j\omega_0) = 8.36\,\Omega + \frac{1}{j\omega_0\,74.21\,\text{pF}}. \tag{6.7a}$$

In order to be resonant at this frequency, the inductance[8] of the tag should be optimized to

$$L_{\text{tag,desired}} = \frac{1}{\omega_0^2\,74.21\,\text{pF}} = 1.86\,\mu\text{H}. \tag{6.7b}$$

This inductance is obtained in the following optimization process by adapting geometrical parameters only.

An additional optimization goal is to simultaneously maximize the quality factor of the coil. As already been discussed in section 3.3.3.1, a maximized Q-factor of the inductor is mandatory for achieving a high power-transfer efficiency and consequently low ohmic losses inside the conductors. Due to the fact that the outer dimensions of the coil are fixed in this example, the remaining parameters to be optimized are w, s and N_{turn}.[9]

Name	Value
C_{Load}	74 pF
R_{Load}	3 kΩ

Table 6.6: Transponder IC input impedance

Instead of applying a global optimization algorithm, the influence of the three parameters w, s and N_{turn} is first analyzed by a parameter sweep which is performed with the LQS-PEEC method.[10] The results are presented in Figure 6.19 in which the intrinsic quality factor of (3.13a) is visualized as a function of the conductor width w and the number of turns N_{turn}. The top and bottom plots show the results for two different spacing values s in order to account for the influence of the conductor spacing on the results.

---

[7]If the system is operated in resonance mode, the working and the resonance frequency coincide.
[8]Here, the inductance includes the capacitive effects with $L_{\text{tag}} = \text{Im}\{Z_{\text{QS}}\}/\omega$ and differs from the MQS inductance L_2 according to Figure 3.12.
[9]One might think of further design parameters to be optimized such as introducing a curvature of the corners or using "tapered spirals" as presented in [52, 124, 125]. For simplicity reasons, these additional design parameters are not considered here.
[10]The settings are according to the fine mesh of Table 6.7.

6.2 Printed Spiral Coil

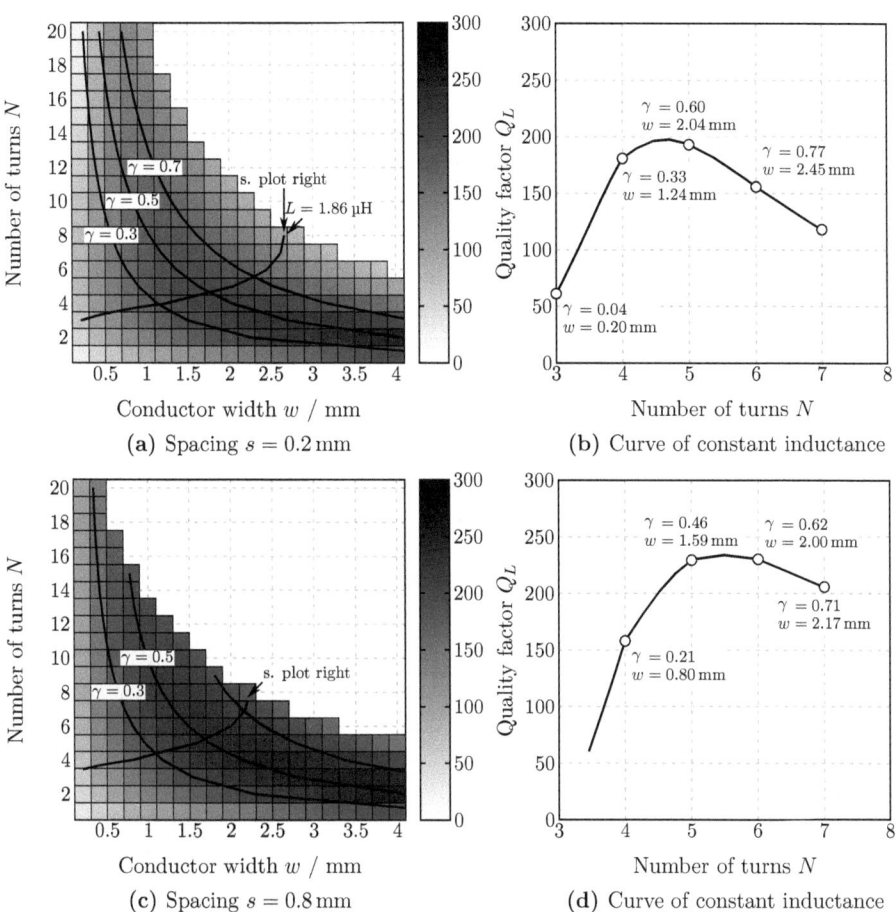

Figure 6.19: Quality factor for different turn numbers, widths and spacings for the test setup according to Table 6.5. The two left figures (a) and (c) show parameter sweeps over w and N_turn for two different spacing values. The top-right areas cannot be evaluated since the available coil size permits too many windings for large trace widths. In order to emphasize this effect, three isolines with a constant fill factor γ from (3.16) are appended to the graphs. It can be seen that a fill factor of approximately 0.5 yields maximum quality factors. In the left two figures, the black isolines of the desired inductance 1.86 µH motivate to optimize the quality factor independently of the inductance. In the right two figures (b) and (d), the curves of constant inductance are presented as a function of the number of windings whereas the fill factors and widths change for each turn number. This allows for concluding that $N_\text{turn} = 5$ and a fill factor of approximately $\gamma = 0.5$ provide optimal results.

Chapter 6 Simulation Results and Measurements

First, the two left figures are analyzed. For very low turn numbers and conductor widths, the quality factor is low since the losses w.r.t. the inductance are high. The Q-factor can either be increased by enlarging the width of the traces (going right) or by increasing the number of turns (going top). This relation can be verified by the fact that the inductance grows with N_{turn}^2 (s. page 28) while the resistance grows with N_{turn} only. In case of spiral coils, this statement can of course be used as a rough estimation only. In the left figures and especially in Figure 6.19c it is visualized that the maximum quality-factor values are obtained in an intermediate region of both design parameters, the width and the number of turns. This fact can be emphasized when plotting lines of constant fill factor γ of (3.16) which expresses the area filled by the conducting material w.r.t. the whole coil area. It can be observed that the Q-factor approximately corresponds to lines with the same fill factor. In this example a fill factor of about $\gamma = 0.5$ provides best results.

Some comments about the parameter settings in the right and top areas of Figure 6.19a and Figure 6.19c should be made. For too large numbers of windings, the capacitive effects increase and therefore the intrinsic quality factor decreases.[11] The other way around, too wide traces and very low turn numbers result in inaccurate results since the increased influence of the conductor-bend effects is not modeled correctly by the simplified corner modeling. When regarding the influence of the spacing s on the results, it can be observed that too low spacings cause increased losses due to the increased proximity-effect losses. On the other hand, too large spacing values result in a non-optimum exploiting of the total available coil area.

Summing up, the quality factor does not have a sharp maximum in the parameter space. Moreover, best results are obtained for intermediate parameter values for which approximately half of the overall coil area is filled with conducting traces. These findings allow for choosing the inductance relatively independent of the quality factor. For this reason, the black lines of the left plots of Figure 6.19 indicate the desired inductance $L_{\text{tag,desired}}$ of (6.7b) in the parameter space. This line is plotted in Figure 6.19b and Figure 6.19d as a function of the number of turns. It is seen from the figures that the inductance can either be reached by applying a low number of turns as well as a low conductor width or, alternatively, by using more turns and an increased conductor width. As expected before, maximum Q-values are obtained for a fill factor of about $\gamma = 0.5$. Since this fill factor is best achieved for $N_{\text{turn}} = 5$, this turn configuration is chosen in the following.

In the next step, the width w and the spacing s are optimized for achieving the desired inductance and a maximum quality factor. In contrast to the previous parameter sweep, the vias and the diagonal return conductor on the bottom layer (s. Figure 6.20) are now considered. The optimization is based on the Nelder-Mead method [126] which minimizes a scalar-valued nonlinear function of multiple real variables without any derivative

---

[11]This is not necessarily the case if a different Q-factor definition is used.

6.2 Printed Spiral Coil

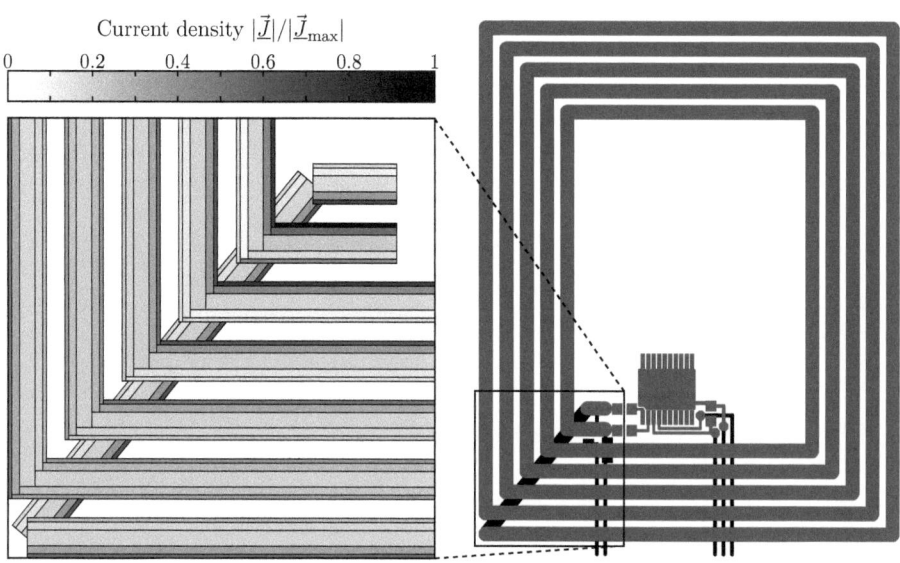

(a) PEEC mesh and current distribution (b) Finalized layout with IC footprint

Figure 6.20: Finalized layout of the optimized PSC with the parameter values from Table 6.5. (b) Fabricated design including the footprint of the IC which can be connected to the antenna via $0\,\Omega$ bridges. The vertical traces on the bottom of the figure are used for measurement purposes (s. Figure 6.24). (a) Excerpt of the LQS-PEEC mesh with $N_w = 5$ and $N_t = 1$. The cells are colored with the current distribution at 1 MHz in order to visualize the influence of the skin and proximity effects. The real discretization is shown in Table 6.7.

information. In this case, the function $f(w,s)$ depends on the variables w and s. It is chosen as
$$f(w,s) = 10\,\frac{|L_{\text{tag,desired}} - L(w,s)|}{L_{\text{tag,desired}}} + \frac{|Q_{\max} - Q(w,s)|}{Q_{\max}}, \quad (6.8)$$
with $Q_{\max} = 300$. The error in inductance is weighted by a factor of 10 since it is considered as more important. By using the start values obtained by the sweep from Figure 6.19d, approximately 10 to 20 iterations are required to obtain a result with an inductance error of less than 0.1 % and a quality factor of $Q = 214$.[12] The optimized values are presented in Table 6.5 in the gray shaded cells while the geometry is visualized in Figure 6.20. The right Figure 6.20b shows the finalized layout including the IC footprint and some additional conductors which can optionally be connected for measuring purposes (s. Figure 6.24). In Figure 6.20a, an exemplary PEEC mesh with a current

---
[12] Figure 6.19 might suggest that a quality factor of up to 240 can be reached. This is not the case because the return conductor has been neglected in the parameter sweep.

Chapter 6 Simulation Results and Measurements

| FEM | | PEEC | | |
| | | | Fine | Coarse |
Name	Value	Name	Value	Value
Tetrahedrons	$2\,10^6$	Current Cells	2 186	301
DoF	$37.6\,10^6$	Charge Cells	560	46
Basis fun.	3rd	Basis fun.	0th	0th
Boundary	electric	Boundary	—	—
Bound. box	200 mm	l_{max} [a]	20 mm	—
Solv. accuracy	10^{-4}	Solv. accuracy	direct	direct
N_w	approx. 8	$N_w, N_t, N_{w,pan}$	13, 3, 5	13, 1, 1
Memory	58.6 GB	Memory	some MB	kB – MB
Overall time[b]	66 h	Overall time	11.5 min	6.7 s

[a]Maximum allowed segment length between two nodes.
[b]13 frequency points are evaluated.

Table 6.7: Solver settings of the full-wave FEM and LQS-PEEC models. For an accurate modeling of the eddy currents, a dense volume mesh is required inside the conductors. In this case, an adapted efficient PEEC mesh with large aspect ratios of the cells and a restriction of the current cells in the direction of the estimated current flow enables a remarkable speedup compared to the general purpose FEM solver.

distribution for $f = 1\,\text{MHz}$ are visualized. Besides the skin effect, also the proximity effect greatly impact the overall current distribution which is highest at the inner side of the innermost winding. The relatively coarse subdivision of the conductors into five segments each is done for visualization aspects only. The concrete mesh settings will be focused on in the next paragraph.

Comparison of FEM and LQS-PEEC Results In order to obtain a reference result for the optimized coil layout displayed in Figure 6.20, a FEM simulation is performed with the settings according to the single-turn coil example of section 6.2.2 and Table 6.7. The results separated by real and imaginary parts are visualized in Figure 6.21a and Figure 6.21b. The overall simulation time for 13 frequency points is about 2.8 days.

In the following, the PEEC simulations are performed with two different mesh settings in which it is distinguished between *fine* and *coarse* simulations. The specific settings are visualized in Table 6.7. The inductive mesh is equivalent to the single-turn coil of the last section which is based on the cross section subdivision of appendix C on page 177 with $\chi = 2$ and the corner modeling via a single node as presented in Figure 4.7c. For this setup, $N_w = 13$ and $N_t = 3$ are obtained for the fine mesh and $N_w = 13$ and $N_t = 1$ for the coarse mesh, respectively. The capacitive mesh is in accordance

6.2 Printed Spiral Coil

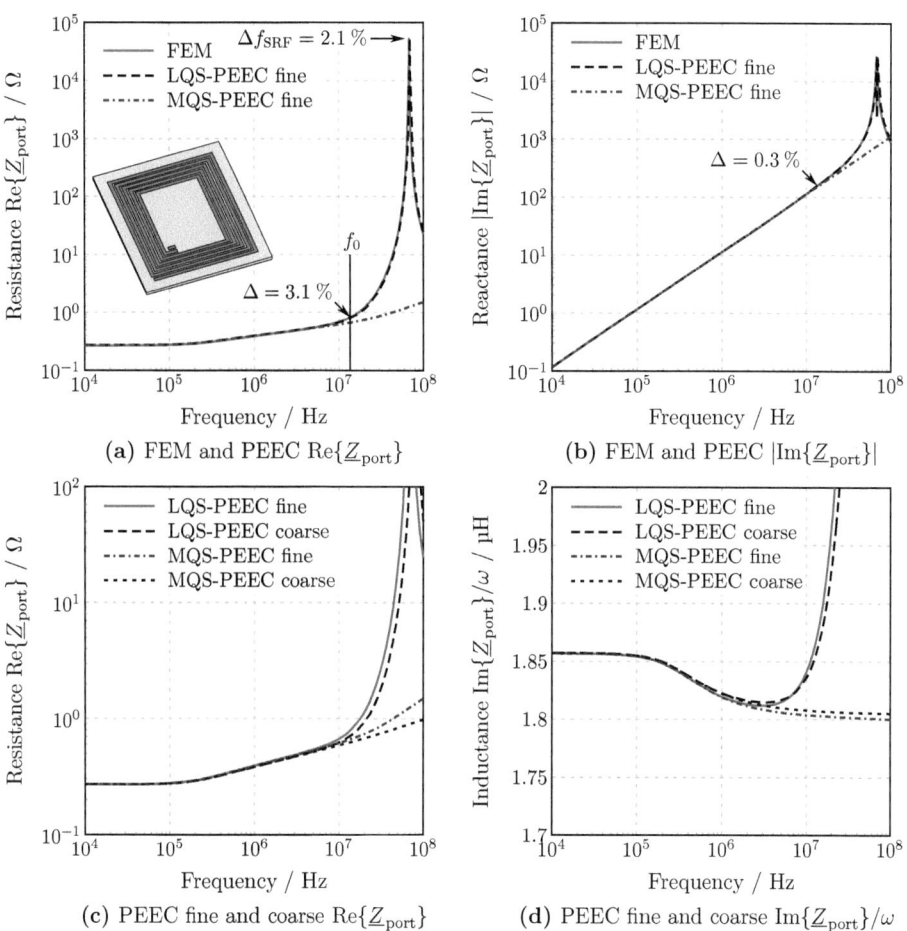

Figure 6.21: Comparison of the FEM and the PEEC results for the optimized PSC with the geometrical parameters of Table 6.5. In (a) and (b), the FEM reference simulation is compared with the fine LQS-PEEC and MQS-PEEC results with the solver settings according to Table 6.7. As already noticed in the previous sections, deviations of a few percent can be observed due to the simplified corner modeling as well as the efficient cross-section modeling. When comparing (a) and (b), the inductance error tends to be smaller than the resistance error. Moreover, the MQS results do not show resonant behavior as expected. This is due to the neglect of the electric energy. In (c) and (d), the influence of the mesh on the results is presented. For initial design purposes, a coarse mesh may be sufficient since the overall simulation time is reduced from 11.5 min to 6.7 s. From (d) it is obvious that capacitive effects are not negligible for frequencies above 10 MHz.

Chapter 6 Simulation Results and Measurements

with section 4.5.3 whereas in the fine simulation, the conductor widths are subdivided by $N_{w,\text{pan}} = 5$ elements each. In addition, the segment length is limited in the fine simulation to $l_{\max} = 20$ mm which allows a better modeling near the SRF. The PEEC simulations are performed by setting up the matrices of partial elements, the nodal connectivity matrix as well as the terminal incidence matrix. Afterwards, (4.22) is solved for the port impedance of the coil. For the MQS-PEEC models, the capacitance matrix is substituted by zeros according to (4.25).

The simulation results of the fine mesh are compared in Figure 6.21a and Figure 6.21b with the FEM results. The results agree up to a few percent which has already been motivated in the previous sections due to the approximative skin-effect modeling and the simplified corner discretization. In addition, a shift in the resonance frequency of about 2.1 % can be observed. This can be explained by the relatively long segment length on the one hand and a neglect of the retardation effects on the other hand. As mentioned before the reason is a different electric energy (cf. footnote 8 on page 52). Nevertheless, the accordance of both FEM and LQS-PEEC results is sufficient for practical applications.

The influence of the PEEC mesh is shown in Figure 6.21c and Figure 6.21d, where the fine and coarse simulations are compared with each other. As can be seen from the figures, the differences grow for increasing frequencies. However, the coarse simulation might be favorable for initial design purposes since the simulation time is reduced from 11.5 min to 6.7 s. A more detailed discussion about the discretization settings can be found in [123].

Equivalent Circuit Model In this paragraph, the parameter values of the reduced equivalent circuit models of section 3.3.2 are derived for the optimized PSC. For the narrowband models according to Figure 3.7, the fine MQS-PEEC and the fine LQS-PEEC models are solved at f_0. This allows for computing the four values R_p, C_p, L_s and R_s of Table 6.8 according to (3.22). It is seen that the MQS inductance L_s is a few percent lower than the LQS-PEEC inductance of the optimization process. This matter of fact can also be verified by Figure 6.21d in which the capacitive influence causes an increase of the inductance. The parameter values of the broadband network model displayed in Figure 3.8 are also specified in Table 6.8. The values are obtained according to the three fitting steps of section 3.3.2.2. First, two MQS-PEEC and two LQS-PEEC evaluations at 1 MHz and 70.27 MHz are performed in order to extract R_p and C_p. Second, two further MQS-PEEC simulations at 10 Hz and 1 THz are carried out for computing R_DC, L_int and L_ext. For the third step, eight additional MQS simulations at logarithmic spaced frequency points from 10 kHz to 100 MHz are evaluated in order to obtain the ladder-model parameters according to Figure 3.8. In this case, a model order of six is chosen as it provides accurate results. More details about choosing the appropriate order are presented in [55].

6.2 Printed Spiral Coil

	Broadband model $f_{SRF} = 69.2\,\text{MHz}$				Narrowband model $f_{SRF} = 70.2\,\text{MHz}$	
Name	Value	Name	Value		Name	Value
R_p	61.3 kΩ[a]	L_{int}	57.8 nH		R_p	340 kΩ
C_p	2.94 pF	L_{ext}	1.80 µH		C_p	2.85 pF
R_{DC}	271 mΩ				R_s	664 mΩ
R_1	193 mΩ	L_1	17.2 nH		L_s	1.80 µH
R_2	177 mΩ	L_2	10.3 nH			
R_3	305 mΩ	L_3	5.94 nH			
R_4	616 mΩ	L_4	2.35 nH			
R_5	524 mΩ	L_5	69.0 pH			
R_6	67.5 mΩ					

[a]The frequency depend resistance $R_{p,\text{freq}}(f) = R_p f_{SRF}/f$ is considered in this example which is 312.8 kΩ at f_0.

Table 6.8: Extracted equivalent circuit parameters of the test coil obtained from the fine PEEC simulations. The network topologies are according to Figure 3.7 and Figure 3.8. A comparison between the full simulations and the reduced models is presented in Figure 6.22. In the ladder model, an order of six is sufficient which can be seen from the small inductance value L_5 compared to L_1 to L_4.

In Figure 6.22, the port impedance of the transponder antenna separated by resistance and inductance is visualized. While the full model consists of about $1.24\,10^6$ circuit elements, the reduced broadband model contains 16 elements only. The narrowband model even further reduces the number of required network elements to four. While the broadband model provides accurate results over the whole considered spectrum, the narrowband model is able to represent the behavior near the working frequency only. The SRF is modeled by both approaches (s. Table 6.8). It tends to be more accurate in the broadband model due to the iterative fitting routine. Despite of this fact, the narrowband model will be used in section 6.3 where an antenna system consisting of two inductively coupled coils is analyzed. This is because the analysis will be performed at a single frequency only.

Measurements In order to verify the previous simulation results, the fabricated PSC with the layout according to Figure 6.20b is measured via an impedance analyzer [127] ranging from 40 Hz to 110 MHz. In particular, the measurements are performed with a pin probe which is connected to the antenna via two pads. Due to the fact that the connector introduces some additional capacitive couplings to the conductors, this effect needs to be de-embedded for a proper comparison with the simulations. For this reason, the SRF is measured contactlessly in a second setup via the concept of the

Chapter 6 Simulation Results and Measurements

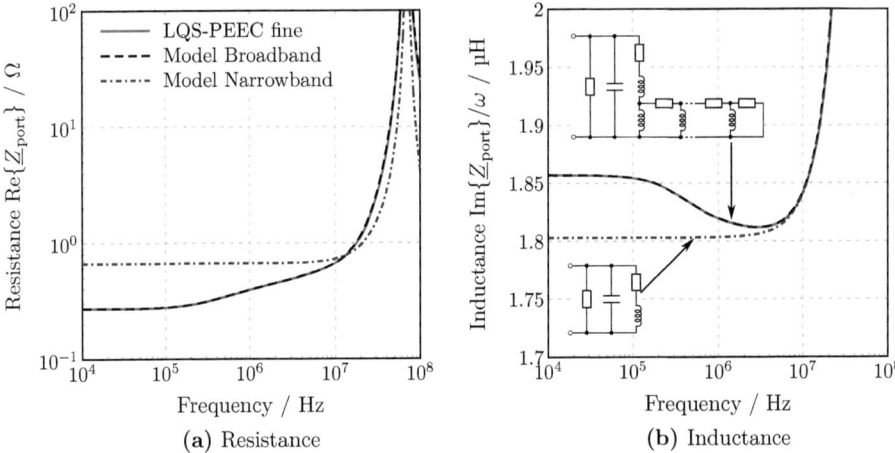

Figure 6.22: Comparison of the full LQS-PEEC results and the reduced macromodels presented in Figure 3.8 and Figure 3.7b. All physical relevant properties such as the SRF and frequency-dependent losses are modeled correctly by the broadband model consisting of 16 lumped RLC elements according to Table 6.8. Considering the narrowband model with 4 circuit elements only, the behavior is not modeled correctly in the whole spectrum. Instead, coincidence is given only near the working frequency as well as the SRF.

reflected impedance (3.34). The measurement is realized by connecting an arbitrarily shaped closed conductor loop to the impedance analyzer, performing a calibration and positioning the PSC at a distance of a few centimeter in front of the loop. At the SRF, the maximum current is induced in the coil. This results in the maximum real part of the reflected impedance (3.34) which can be measured by the impedance analyzer. By evaluating the bandwidth of the reflected impedance curve, the quality factor can be computed via (3.14b).

Name	Value
$R_{\text{s,PinProbe}}$	287.6 Ω
$C_{\text{s,PinProbe}}$	613.4 fF
$R_{\text{s,Footprint}}$	71.1 Ω
$C_{\text{s,Footprint}}$	523.5 fF

Table 6.9: De-embedding

When comparing the two direct and contactless measurements near the SRF, the influence of the pin probe can be quantified as a series connection of $R_{\text{s,PinProbe}}$ and $C_{\text{s,PinProbe}}$ with the parameter values of Table 6.9. In the following, this impedance is subtracted from the pin-probe measurements. The results are plotted in Figure 6.23 in terms of the resistance, inductance, quality factor and dissipation factor. In order to compare the results with the FEM and LQS-PEEC simulations, a further de-embedding procedure of the simulation results is performed to account for the differences of the simplified model presented in

6.2 Printed Spiral Coil

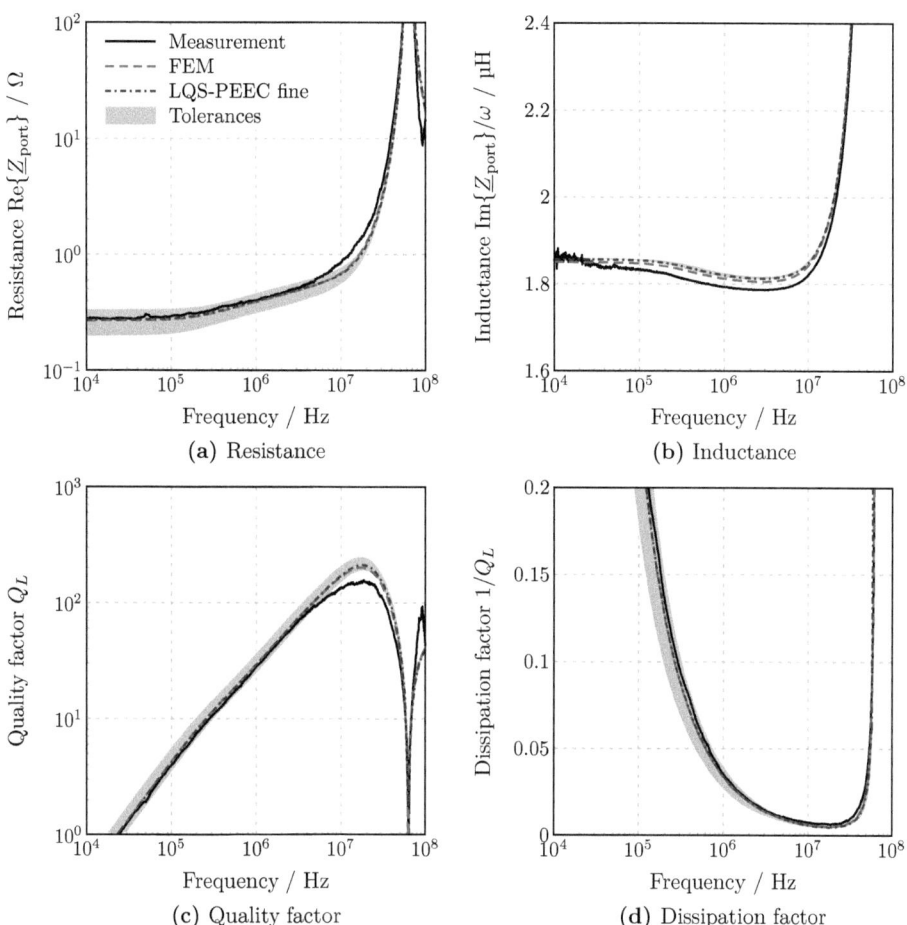

Figure 6.23: Comparison of the measurements with the PEEC simulation as well as the FEM reference solution for the optimized coil geometry displayed in Figure 6.20. In order to better compare the different results, the quality factor and the dissipation factor are visualized next to the resistance and the inductance. The different curves have been de-embedded with serial RC elements of Table 6.9 in order to provide comparable conditions for all results. The errors of the measurements w.r.t. the FEM results at 13.56 MHz are 30.1 % at the resistance and 1.3 % at the inductance. These deviations do not exceed the measurement tolerances of the impedance analyzer [127]. Besides the measurement and the simulation curves, a sensitivity analysis of the LQS-PEEC results has been appended (gray shaded areas) in which the worst case values with the basic tolerances of Table 6.5 have been taken as a basis.

Figure 6.20a[13] and the fabricated layout according to Figure 6.20b. Again, the influence is mainly of capacitive nature since the IC footprint and the measuring traces are electrically not connected to the antenna. The influence of these additional conductors is de-embedded by simulating two different FEM models including and excluding these components. For a better comparison, the mesh settings are chosen as equal as possible in both models. In the frequency range near the SRFs of both results, a parallel RLC resonance circuit is fitted for each model whereas the inductances are chosen identically in both cases. By evaluating the differences of the obtained capacitances and resistances, an equivalent circuit model is extracted, accounting for the influence of the additional components as a series connection of $R_{\text{s,Footprint}}$ and $C_{\text{s,Footprint}}$. As a result, the values of Table 6.9 are obtained.

The results of the above de-embedding procedure applied to the FEM and LQS-PEEC models are appended to Figure 6.23. The measurement errors at $f_0 = 13.56$ MHz w.r.t. the FEM results are 30.1 % for the resistance and 1.3 % for the inductance. These errors coincide with the accuracy range of the impedance analyzer [127] which allows for concluding a good accordance to the measurements over the considered frequency spectrum.

Sensitivity Analysis At the end of this section, a sensitivity analysis is performed for the PSC of Table 6.5 in which the manufacturing tolerances of the particular design parameters are included. Caused by the fabrication process, six tolerance-associated parameters exist in this setup, i.e. three geometrical parameters w, t and h as well as three material properties ε_r, $\tan \delta$ and κ. When the gradient information of the port impedance w.r.t. these parameters is known, a linearization of the actual design can be set up. This allows the analysis of the design parameter influences on the system behavior.

In order to obtain the derivative information of the port impedance, the adjoint sensitivity equation (5.6b) is applied to the MNA single port system (4.21). Due to the symmetrical character of the system, the original system solution and the adjoint counterpart are identical. Consequently, the initial system needs to be solved just once. The remaining part of the sensitivity analysis concentrates on the computation of the system matrix derivatives w.r.t. to the six design parameters. This is achieved by using a forward FD approximation according to (5.7) or, more precisely, (6.4b) in case of the partial inductances. In the FD approximations, a relative parameter perturbation of 10^{-3} is used to approximate the derivatives of the matrix elements in a reasonable manner (s. Figure 6.13). Prior to that, an element perturbation algorithm checks which matrix entries have to be recomputed. This approach is time-saving since generally not all elements are perturbed for each geometrical variation.

---

[13] In this model, the IC footprint as well as the measuring conductors are not considered.

As an example, if the thickness h of the substrate is varied, the relative position and orientation of the elements on the top and bottom layers do not change. From this follows that a large area of the system matrix derivatives is zero and consequently does not need to be recomputed. Obviously, the amount of perturbed elements may differ for each analyzed parameter. In the considered test setup, the worst case values are the conductor widths and thicknesses, respectively. This is due to the fact that in the uniform perturbation approach according to Figure 5.2b, all elements are scaled and shifted. Thus, a re-computation of the same number of elements as required for the original system is demanded.[14] More precise, the variation of the substrate thickness h requires about 6.8 % of the original matrix fill while the material properties ε_r and $\tan\delta$ only demand a re-calculation of the capacitive elements due to the modified Green's function of (A.12). The cheapest design parameter is the conductivity κ since it influences only the derivatives of the partial resistances.

After computing the port impedance derivatives w.r.t. to the different design parameters, the worst case tolerances of Table 6.5 are used in order to obtain the limits of the impedances by means of a linearization. The results are included in Figure 6.23 in which the gray shaded areas indicate the borders in which the actual curve could be located. It is observed that the measurements do not exceed the tolerance limit for a broad frequency range. The regions in which the curves exceed the boundaries can be explained by measurement inaccurateness [127].

6.3 Inductively Coupled Antenna System

In the last part of the results chapter, the system behavior of an IPT antenna system consisting of the multi-turn PSC of the previous section and a single-turn reader antenna is analyzed. A spatial sweep of the relative antenna arrangement is performed. For each position, the measured and simulated results are compared. Since the antenna system is compatible with the standard High Frequency (HF) RFID technique, not only voltages are considered but also the wireless communication link is tested in terms of identifying the transponder. Because the simulations are based on the reduced circuit models of section 3.3, the simulations can be performed in milliseconds, hence allowing for a fast and precise forecast of the readout range.

6.3.1 Setup of the RFID Antenna System

Although the primary goal of RFID systems is to identify transponders by wireless data transmission, the contactless power transfer is an important property for many low-cost systems since a battery at the transponder circuit must be avoided. In Figure 6.24, the considered antenna setup is visualized. The system is operated in the HF band and

---

[14]The ILC could again reduce the number of required element couplings but is not applied here.

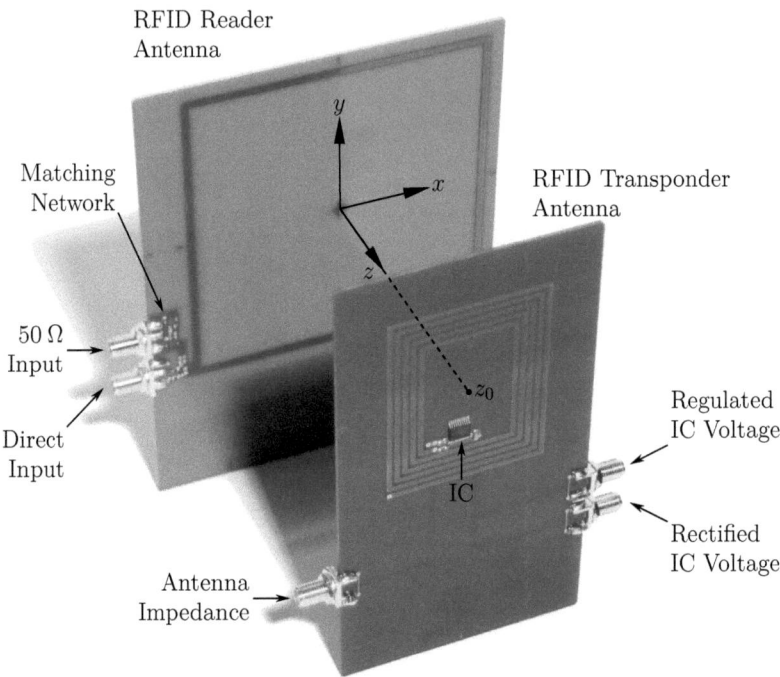

Figure 6.24: Measurement setup of the IPT system in which the multi-turn transponder antenna of the last section is located in front of a square single-turn reader antenna with a side length of 100 mm. The reader antenna can be operated by either attaching a 50 Ω source or a source connected directly to the coil. The transponder IC can be read via RFID. The possibility to measure the rectified and regulated coil voltages is also provided. For flexibility and cost reasons, the antennas are fabricated in standard PCB technology.

in particular at $f_0 = 13.56$ MHz which is part of an ISM band. The transponder circuit is composed of the optimized five-turn PSC of section 6.2.3 as well as the transponder IC with the parameters values of Table 6.6. The transponder is mutually coupled with a single-turn square reader antenna having a side length of 100 mm and a trace width of 2 mm. The material properties of the reader antenna are equivalent to Table 6.5 as the same technology is used for both coils.

In order to extract the circuit properties of the reader coil which are needed for the circuit simulations, two different numerical PEEC simulations are performed at 13.56 MHz, the LQS-PEEC as well as the MQS-PEEC methods. In both models, the mesh settings are chosen equivalently with the constraints of the multi-turn coil as

6.3 Inductively Coupled Antenna System

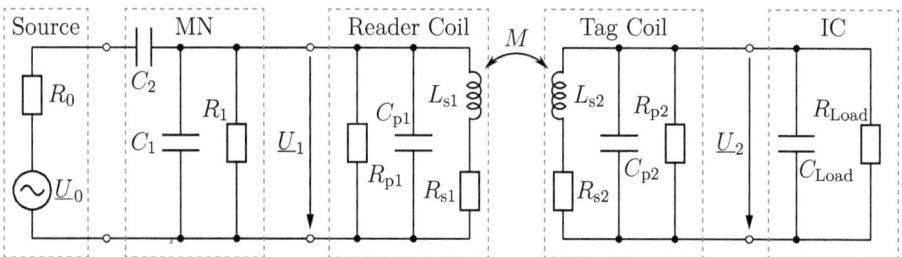

Figure 6.25: Equivalent circuit description of the RFID test setup. The parameter values of the transponder circuit can be found in Table 6.8 and Table 6.6 whereas the reader circuit settings are presented in Table 6.10. The mutual inductance is a function of the spatial arrangement of the coils in this analysis. Two different voltages $\underline{U}_1$ and $\underline{U}_2$ have been introduced for measuring purposes.

Name	Value	Name	Value
$\underline{U}_0$	7.4 V	R_{p1}	1.63 MΩ
R_0	50 Ω	C_{p1}	1.01 pF
R_1	1 kΩ	R_{s1}	201 mΩ
C_1	331 pF	L_{s1}	355 nH
C_2	59.8 pF		

Table 6.10: Parameter values of the equivalent circuit of the reader antenna unit according to Figure 6.25 consisting of the source, the matching network and the coil. As before, the voltage source is characterized by its RMS value which is capable of powering a 50 Ω load with 274 mW.

discussed above. The results of the two different simulations allow the extraction of the narrowband equivalent circuit model displayed in Figure 3.7b with the parameter values being presented in Table 6.10.

As can be seen from Figure 6.24, the RFID antenna system can be powered via two different inputs whereas either of them can be connected to the antenna. The first possibility is to directly connect the source to the antenna. This is done in section 6.3.3 where the impedance analyzer [127] is used to measure the reflected transponder impedance. Alternatively, a matching network can be connected in between the antenna and the source in order to provide an input impedance with a real value of 50 Ω. This is convenient for connecting the antenna to transceiver units which are often designed to power a 50 Ω load. Results of this approach will be presented in section 6.3.4.

The matching network used throughout this section is composed of one resistor and two variable capacitors which can be optimized in order to obtain an input impedance of 50 Ω. The resistor reduces the quality factor of the resonance circuit. Consequently, it ensures a more robust behavior for varying coupling conditions in which the reflected impedance of (3.31) changes. The matching network design is discussed more detailed in [70]; the actual realization is visualized in Figure 6.25 with the parameter values presented in Table 6.10.

Chapter 6 Simulation Results and Measurements

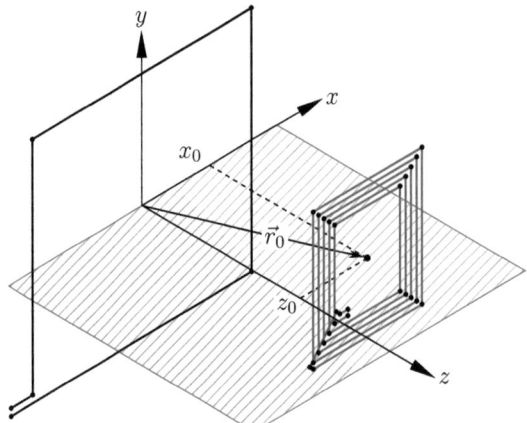

Figure 6.26: Filamentary model of the setup presented in Figure 6.24 for computing the mutual inductance between both PSCs. The square reader antenna is composed of 6 nodes while the transponder antenna is modeled by 26 nodes. In the test setup according to Figure 6.29, the transponder is swept over the gray highlighted area. In each position, the mutual inductance computation requires a few milliseconds only.

The equivalent network description of the overall setup is visualized in Figure 6.25. The source is modeled as an ideal voltage source of 7.4 V RMS connected in series with the internal resistance of 50 Ω, thus allowing to power a 50 Ω load with 274 mW. Although not shown in Figure 6.25, the mutual inductance $M(\vec{r}_0)$ is a function of the spatial separation of both antennas, indicated by the position of the center point of the transponder antenna $\vec{r}_0$ relatively to the reader antenna (s. Figure 6.26). An arbitrary orientation of the two antennas as depicted in Figure 3.1 is not considered in this section. This is because a positioning robot is used which is not able to rotate the attached antennas. Nevertheless, the mutual inductance extraction technique of section 4.5.4 works for arbitrary 3D setups.

6.3.2 Mutual Inductance Computation

In this section, the PEEC settings of the upcoming simulations are briefly presented. As already motivated before, the aim is to decouple the simulation of the individual antennas from the mutual antenna coupling while recovering the whole system behavior in the circuit domain according to Figure 6.25. An alternative would be to repeatedly simulate the complete antenna system for each spatial position. However, this approach is abandoned here since it is much more time consuming. This is mainly caused by the increased system matrix consisting of both antenna models for which parts must be recomputed for each spatial orientation. More details of comparing the full model and the reduced network model can be found in [55] where an excellent agreement between both approaches is shown.

After extracting the macromodels of the individual antennas, the only remaining unknown parameter in the network model displayed in Figure 6.25 is the mutual inductance

$M(\vec{r}_0)$. According to the explanation of section 4.5.4, the mutual inductance between both coils is evaluated by the MQS-PEEC method. For this purpose, the minimal mesh settings are used in which each straight conductor segment is modeled by a single current cell only. The actual discretization of the test setup is visualized in Figure 6.26. The reader coil is modeled by $N_{n1} = 6$ nodes and $N_{b1} = 5$ branches. The transponder coil is modeled by $N_{n2} = 26$ nodes and $N_{b2} = 25$ branches, respectively. In order to compute the mutual inductance via (4.35), the computation of $N_{b1} N_{b2} = 125$ mutual inductances is done by the filamentary approach (s. appendix A.1 on page 160), lasting only about 13 ms on a standard desktop computer.

6.3.3 Measurements of the Reader Antenna Input Impedance

In the first measurement setup, the impedance analyzer [127] is directly connected to the reader antenna whereas the matching network is detached from the circuit. In Figure 6.25, the measuring point is indicated by $\underline{U}_1$. Both coils are mounted on a positioning robot which is able to adjust the position of both coils with an accuracy in the sub-millimeter range. In this setup, the positioning robot is used to vary the relative distance of both coils along the z-axis (s. Figure 6.24). In each position z_0, the impedance analyzer creates a voltage signal of 1 V RMS and evaluates the impedance. The results are visualized in Figure 6.27a for the real part of the input impedance. For each spatial separation, the frequency is swept from 12 MHz to 15 MHz and the impedance is measured. As can be seen from the figure, the influence of the reflected impedance of (3.34) is maximal at the resonance frequency of the transponder. At this frequency, the transferred effective power is maximum.

Two more interesting properties of Figure 6.27a are the observable resonance frequency shift towards higher frequencies and the violation of the resonance-curve symmetry when the coupling is decreased. These facts can be explained by the nonlinear behavior of the rectifying circuit diodes. This results in a voltage-dependent input capacitance and consequently in a detuning of the resonance circuit.

For the PEEC simulations, the mutual inductance M is computed for each z_0. At the same time, the system visualized in Figure 6.25 is solved for the reader antenna input impedance without the matching network. The results are presented in Figure 6.27b from which a good agreement with the measured values can be observed. In contrast to the measurements, the nonlinear behavior is not visible in the simulations since the IC is modeled by a single impedance only.

6.3.4 Measurements of the Data and Energy Transmission

In the last study, the matching network as well as a commercial transceiver unit are connected to the reader antenna. The parameters of the transceiver unit are found from measuring the power delivered to a 50 Ω load which is 274 mW. In order to transfer

Chapter 6 Simulation Results and Measurements

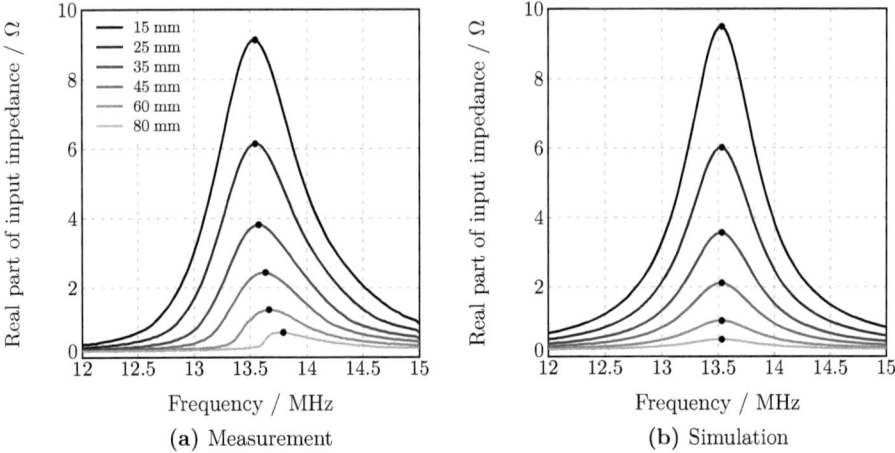

Figure 6.27: Measured and simulated reader antenna input impedance for six different antenna separations z_0. While the resonance frequency in the simulations (b) is almost independent of the coupling factor, the measurements (a) show nonlinear behavior of the IC input capacitance which is mainly caused by the diodes of the rectifying circuit. The black dots indicate the maximum value of each curve.

data over the wireless link, the carrier signal is superimposed by the data signal. A detailed description of the bidirectional data transmission can be found in [9]. Here, it is detected only whether a communication in terms of requesting and responding the identification number of the transponder has been successful.

In Figure 6.28, the measured powering and readout range are compared for varying z_0 ($x_0 = 0$). For the comparison with the simulated data, the computed transponder voltage $|\underline{U}_2|$ according to Figure 6.25 is appended to the graph. When inspecting the measured DC voltage which is regulated to 3 V, a powering range of up to $z_0 = 145$ mm can be concluded. For comparing the measured and the simulated voltages for higher distances, 0.3 V are subtracted from the simulated voltage accounting for losses in the regulator circuit. Due to the fact that the transponder is modeled in the network description by a single impedance only, the simulated voltage $|\underline{U}_2|$ is not limited to any maximum value and consequently increases for decreasing distances.

For the data transmission, the reader aims to set up a bidirectional communication link for each spatial position. In order to reduce noise and to obtain a smoother transition, the data transmission is repeated five times for each position while the number of successful identifications is saved. This allows the conclusion of a readout range from 15 mm to approximately 130 mm.

6.3 Inductively Coupled Antenna System

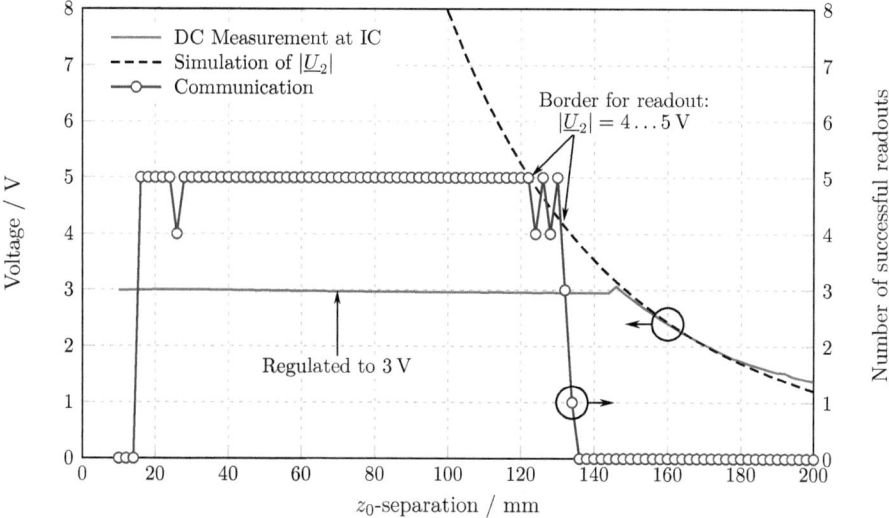

Figure 6.28: Powering and readout range of the test setup for a spatial sweep from 10 mm to 200 mm. The measured and simulated IC voltages exceed 3 V for distances of up to 145 mm. The voltage of 0.3 V has been subtracted from the simulated effective value of the voltage in order to account for the losses in the regulating circuit. For each spatial position, the reader aims to set up a bidirectional data link for five times. On the right y-axis, the number of successful readouts is shown. This allows for concluding a communication range from 15 mm to approximately 130 mm.

The malfunction for too low distances can be explained by the very high signal levels on the one hand and a detuning of the reader circuit caused by large reflected impedance values on the other hand [70]. For too large distances, the data signal level is too low for a proper separation from the noise level. When comparing the maximum readout range with the powering range, the deviation is of about 10 mm to 15 mm.

In order to forecast the readout range for more complex spatial orientations, the positioning robot sweeps the transponder in two dimensions as shown in the gray marked area according to Figure 6.26. The results are presented in Figure 6.29. Again, the data transmission is repeated five times for each position. The number of successful readouts is indicated by a different color in the figure. It can be seen that for too close proximity, the data transmission does not work properly. The maximum readout range is obtained for $x_0 = 0$ where both coils are centered.

When regarding the coupling conditions for low separations in z-direction and about $x_0 = \pm 60$ mm in x-direction, a region of zero coupling can be observed. In this region, no inductive powering of the transponder is possible although both coils are located

Chapter 6 Simulation Results and Measurements

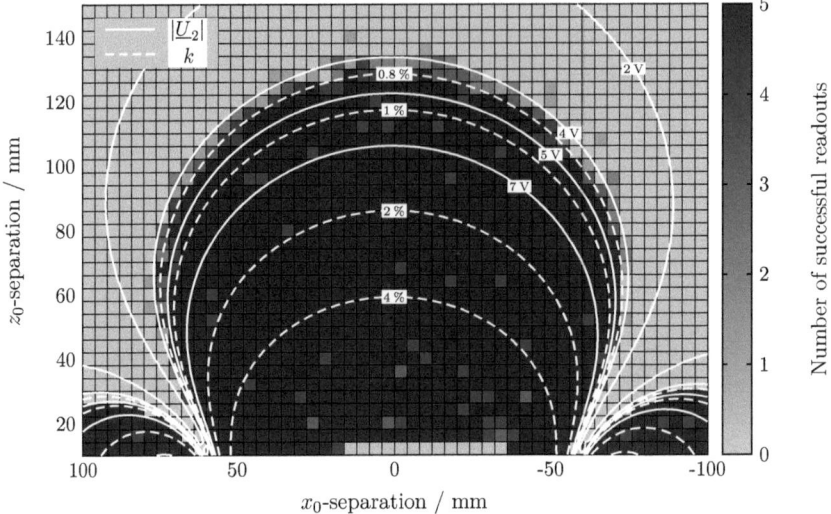

Figure 6.29: Comparison of the measured readout range with the simulated coupling factor and transponder voltage for a 2D spatial sweep as shown in the gray highlighted area as presented in Figure 6.26. As already expected from Figure 6.28, the border of functionality is lying in between 4 V to 5 V of the simulated voltage $|\underline{U}_2|$. Alternatively, a critical coupling factor of about 0.8 % can be concluded which allows for an easy forecast of the working range for any 3D spatial orientation. The time needed to compute the coupling factors for the 1 750 sweep points of this example is about 30 s.

in a very close proximity. This behavior can be explained by the mutual inductance concept based on closed current loops as visualized in Figure 2.3. The magnetic flux density generated by the reader coil changes its direction inside the area bounded by the transponder coil since the coils partially overlap each other. Hence, an overall magnetic flux of zero is obtained and consequently the mutual inductance is zero, too.

Besides the measured communication link of the reader-transponder arrangement, isolines of constant simulated voltage $|\underline{U}_2|$ from Figure 6.25 are appended to Figure 6.29. As a result of Figure 6.28, the border of the readout ranges from $|\underline{U}_2| = 4$ V to $|\underline{U}_2| = 5$ V. This border is confirmed in Figure 6.29.

For practical applications it is desirable to determine a critical coupling factor k (2.48) which sets the border of proper operation. When overlaying different isolines of constant coupling factor to Figure 6.29, the critical coupling factor can be determined to be of about $k_\text{critical} = 0.8$ %. It should be mentioned again that the computation of the mutual inductance between the two PSCs via (4.35) is not limited to parallel arrangements. Instead, it works for any 3D orientation of the coils. The time required to compute

10 000 locations is approximately 2 min for the filamentary setup (cf. Figure 6.26) on a typical desktop computer.

Chapter 6 Simulation Results and Measurements

Chapter 7
Summary and Outlook

In this thesis, inductively coupled antenna systems have been analyzed and designed via a combination of numerical simulations and a network description based on the concept of mutually coupled inductances. The numerical simulations have been performed via a Partial Element Equivalent Circuit (PEEC) solver that was developed and implemented especially for this thesis. The new PEEC solver combines the Lorenz-Quasi-Static (LQS) and Magneto-Quasi-Static (MQS) assumptions and makes use of specialized mesh settings. Hereby, the physical relevant properties of the individual antennas such as frequency-dependent inductance, skin and proximity effects as well as parasitic capacitance can be modeled in seconds to minutes while the discretization errors typically do not exceed a few percent. The obtained results have been used to extract reduced macromodels of the individual antennas for the inclusion in the overall network model. The interaction of the coils has been accounted for via a mutual inductance extraction technique based on a filamentary discretization of the MQS-PEEC approach allowing for spatial parameter sweeps in milliseconds. The macromodels of the coils have been combined with the mutual inductance to an equivalent circuit to be solved in either time- or frequency domain. The proposed approach has been tested with a typical Radio Frequency Identification (RFID) antenna setup. The results have been compared with Finite Element Method (FEM) simulations as well as measurements and a reasonable agreement has been shown. In addition, different concepts for integrating the adjoint sensitivity analysis into the PEEC method have been analyzed and confirmed via exemplary setups.

Contributions of this work

In the following, the particular achievements of this thesis are summarized:

A new LQS approximation of the Maxwell's equations has been motivated and derived. The LQS formulation can be regarded as an intermediate approach between full-wave and MQS or Electro-Quasi-Static (EQS) and is applicable to formulations based on the electric scalar and magnetic vector potentials. The results of the LQS formulation are not novel in terms of practical applications because the same formulation is used by other authors under different names such as Electro-Magneto-Quasi-Static (EMQS) or simply

Chapter 7 Summary and Outlook

Quasi-Static (QS). While these formulations are based on an explicit assumed infinite speed of light in order to get rid of the retardation effects, the new LQS formulation does not need this assumption. Instead, it is consistent with a modified version of the Maxwell's equations and consequently provides a more detailed insight into the underlying fundamentals. Moreover, a slight modification allows for the formulation of the LQS system as a combination of two decoupled electrostatic and MQS systems which are reunited via a joint fulfillment of the continuity equation.

A second novelty of this work is to apply various quasi-stationary approaches in a conjoint simulation and to use the different results in order to extract important physical system properties. This is different to the traditional way in which first the appropriate model is determined and afterwards the simulations are run on this specific model in order to forecast the system behavior. In the context of this work, the LQS-PEEC and MQS-PEEC approaches are simulated together whereas the inductive mesh is chosen identical in both formulations. This only results in a small overhead but allows for the obtaining of different results, one accounting for the capacitive influence and one excluding the same. Hereby, an added value is created, e. g. for extracting macromodels of the antennas.

The extraction of different reduced equivalent circuit models of inductors in both narrowband and broadband regimes is a third main aspect of this work. Although applied to Printed Spiral Coils (PSCs), the approach can also be utilized to other linear passive devices in which the magnetic energy dominates at low frequencies. In particular, the above mentioned combined MQS and LQS simulations at different frequencies have been used to extract reduced network models of PSCs allowing for fast circuit simulations in both time and frequency domain. The benefit is the fact that the models consist of physically motivated circuit elements, thus providing a smart integration of the antenna models into system engineering processes. Results of a test scenario have shown a close agreement between full PEEC simulations and the reduced broadband model over multiple decades by using ten to twenty network parameters for the latter approach only.

The specialized PEEC mesh settings for Inductive Power Transfer (IPT) antenna systems are a further essential feature of the presented modeling procedure as they allow for fast and accurate simulations of the individual coils. This is especially the case if antenna designs with long and thin conductors are being used. In particular, the modeling of rectangular conductor bends has been focused on whereas a 2D approach and two different 1D simplifications have been compared in terms of accuracy and effort. In another case, in which the mutual inductance of different coils is extracted, a coarse filamentary mesh has been used for the MQS-PEEC method which avoids solving a system of equations and has been shown to be equivalent to the Greenhouse method.

The adjoint sensitivity analysis with the main focus on skin-effect problems provides another main point of this work. A method for computing the exact derivatives of

PEEC partial network elements has been derived which can be used for benchmarking purposes, e. g. when applying Finite Difference (FD) approximations. A novel Inner-Layer Concept (ILC) has been introduced which can be used to reduce the amount of element interactions when building the matrix of partial derivatives.

Concerning the IPT system design, a complete analysis has been presented in the network domain by combining the concept of mutually coupled inductances with accurate macromodels consisting of lumped circuit elements. An approach for optimizing the entire system in terms of efficiency maximization and field-emission minimization has been derived for different matching network topologies. In this context, an expression for the reflected impedance of an inductively coupled power receiver at the transmitter equivalent circuit by means of the frequency detuning has been introduced.

Two further contributions of this work include the presentation of an analytical Direct Current (DC) resistance correction term for rectangular conductor bends with different conductor widths as well as a comparison of different analytical approaches for determining the Alternating Current (AC) resistance of an infinitely long conductor with a rectangular cross section.

The measurements of the RFID antenna systems should also be emphasized. They were carried out in the laboratory of the department Advanced System Engineering (ASE) of the Fraunhofer ENAS in Paderborn. Some equipment of the Sensor Technology Department at the University of Paderborn was also used. In particular, the PSC geometries have been optimized whereas the resulting layouts have been created with a Printed Circuit Board (PCB) layout software. The fabricated PCBs were then measured with an impedance analyzer and a positioning robot was utilized to precisely account for the mutual antenna interactions. The controlling of the robot as well as the readout of the measurement equipment were automated.

Outlook

Although the thesis provides a completed analysis approach for the simulation of IPT antenna systems, some further research aspects and improvements have not been addressed.

There remain possible enhancements of the PEEC solver in terms of code optimization and parallelization. In addition, magnetic materials have not been accounted for although they can improve the system behavior in various practical applications due to their possibility to influence and direct magnetic fields. Non-orthogonal PEEC elements have not been addressed either, even though such elements allow for increasing the flexibility of modeling more complex geometries. Furthermore, the applicability of acceleration techniques such as the Fast Multipole Method (FMM) to PEEC systems with long and thin mesh cells can be studied in more detail.

Chapter 7 Summary and Outlook

A further research aspect that remains unaddressed includes the more thorough analysis of the introduced LQS formulation, especially in terms of comparing LQS-PEEC results with full-wave PEEC results for identical discretization settings. This would allow for quantifying the approximation errors more precisely. In addition, enhanced investigations on the borders of functionality for the different LQS and MQS approaches could be accomplished.

Concerning the reduced network description, one improvement could be to include mutual capacitive cross coupling effects of different antennas in the network model. Also, a network ladder model could be introduced in order to account for frequency-dependent dielectric losses by means of lumped network elements.

Appendix A
Partial Network Elements

In this appendix, analytical solutions to the PEEC partial network integrals are presented for some basic geometrical arrangements. Since the computation of the partial resistances is very simple, the following considerations concentrate on the partial inductances and coefficients of potential whereas only the non-retarded integrals are addressed. Besides the presentation of various expressions for the partial inductances, a closed-form solution for the derivatives of the inductances w. r. t. the shape parameters will be outlined since they are needed for the sensitivity analysis. For the coefficients of potential, equations are presented for the free space case as well as a two layer dielectric substrate. It should be mentioned that the symbols used in this appendix do not coincide with the former chapters in all cases. Instead, the notation is done in accordance with the original publications while the meaning of the individual symbols is explained in numerous figures.

A.1 Partial Inductances

The partial inductance of two conductors is generally defined by the double volume integral as

$$L_{mn} = \frac{\mu_0 \vec{e}_m \cdot \vec{e}_n}{4\pi \, A_m A_n} \int\limits_{V_m} \int\limits_{V_n''} \frac{1}{|\vec{r} - \vec{r}''|} \, \mathrm{d}V' \, \mathrm{d}V, \tag{A.1}$$

which is a repeat of (4.12a). Besides the consistency with the general expression (2.45) for homogeneous current distributions, the above equation can also be interpreted by means of the inductance concept based on closed conductor loops in an illustrative manner [72].

Since an analytical solution of (A.1) for two arbitrarily shaped conductors with any desired position and orientation in space is difficult if not impossible to obtain, either numerical or approximative expressions can be applied, e. g. [42, 128, 129]. On the other hand, for special arrangements such as straight parallel conductors with rectan-

Appendix A Partial Network Elements

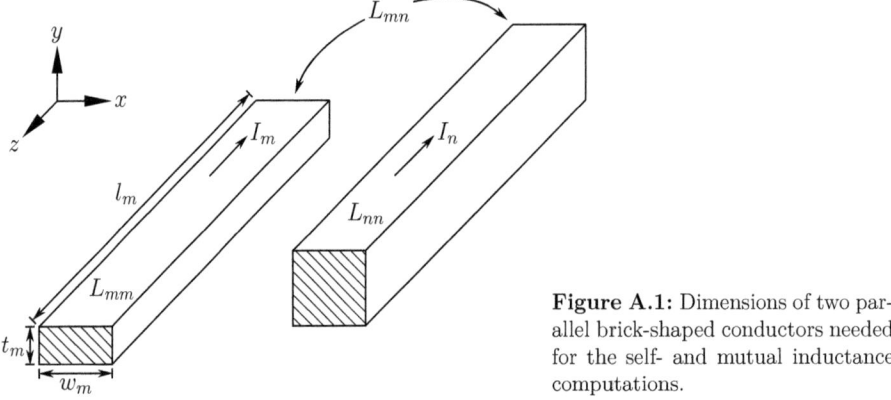

Figure A.1: Dimensions of two parallel brick-shaped conductors needed for the self- and mutual inductance computations.

gular cross sections,[1] closed-form expressions have been found by different authors in the past. In particular, the analytical solution to the six-fold integral (A.1) has been presented by Hoer and Love in 1965 [130] for two arbitrarily positioned parallel brick-shaped conductors as presented in Figure A.1. In 2003, Zhong and Koh [131] derived a numerically more stable formula for the same constraints which will be used in the following.

Due to the fact that the closed-form solution is arduous, several simplifications may be applied for specific geometries. If the conductor thickness is small compared to the width and the length, (A.1) can be reduced to a four-fold integral as presented in [132], for instance. A further simplification is to use a filamentary approach which reduces (A.1) to a double line integral which results in Neumann's formula,[2] e.g. [131]. In 1946, Grover [54] presented a multitude of expressions for Neumann's formula for different filament setups, releasing the parallel precondition. In [42], some rules are presented for choosing the appropriate evaluation technique for different geometries with a special focus on large distance approximations.

In the following paragraphs, the expressions for different practical cases are repeated for completeness reasons.

Self-Inductance of a Rectangular Conductors The evaluation of the self-inductance of a single conductor can be regarded as a special case of (A.1) in which both volumes are identically. If the conductor has the dimensions of conductor m in Figure A.1, the closed-form solution can be represented as shown in the expression of

---

[1] Straight conductors with rectangular cross sections as presented in Figure A.1 are often denoted as bars or bricks.
[2] In this case, the conductors are modeled as infinitely thin and are referred to as filaments.

A.1 Partial Inductances

$$\begin{aligned}
L_{mm} = l_m \frac{2\mu_0}{\pi} \Bigg(\frac{1}{4} &\Bigg[\frac{1}{\tilde{w}} \operatorname{arsinh}\left(\frac{\tilde{w}}{\alpha_t}\right) + \frac{1}{\tilde{t}} \operatorname{arsinh}\left(\frac{\tilde{t}}{\alpha_w}\right) + \operatorname{arsinh}\left(\frac{1}{r}\right) \\
&+ \frac{1}{24}\Bigg[\frac{\tilde{t}^2}{\tilde{w}} \operatorname{arsinh}\left(\frac{\tilde{w}}{\tilde{t}\alpha_t(r+\alpha_r)}\right) + \frac{\tilde{w}^2}{\tilde{t}} \operatorname{arsinh}\left(\frac{\tilde{t}}{\tilde{w}\alpha_w(r+\alpha_r)}\right) \\
&+ \frac{\tilde{t}^2}{\tilde{w}^2} \operatorname{arsinh}\left(\frac{\tilde{w}^2}{\tilde{t}r(\alpha_t+\alpha_r)}\right) + \frac{\tilde{w}^2}{\tilde{t}^2} \operatorname{arsinh}\left(\frac{\tilde{t}^2}{\tilde{w}r(\alpha_w+\alpha_r)}\right) \\
&+ \frac{1}{\tilde{w}\tilde{t}^2} \operatorname{arsinh}\left(\frac{\tilde{w}\tilde{t}^2}{\alpha_t(\alpha_w+\alpha_r)}\right) + \frac{1}{\tilde{t}\tilde{w}^2} \operatorname{arsinh}\left(\frac{\tilde{t}\tilde{w}^2}{\alpha_w(\alpha_t+\alpha_r)}\right) \Bigg] \\
&- \frac{1}{6}\Bigg[\frac{1}{\tilde{w}\tilde{t}} \arctan\left(\frac{\tilde{w}\tilde{t}}{\alpha_r}\right) + \frac{\tilde{t}}{\tilde{w}} \arctan\left(\frac{\tilde{w}}{\tilde{t}\alpha_r}\right) + \frac{\tilde{w}}{\tilde{t}} \arctan\left(\frac{\tilde{t}}{\tilde{w}\alpha_r}\right) \Bigg] \\
&- \frac{1}{60}\Bigg[\frac{(\alpha_r+r+\tilde{t}+\alpha_t)\tilde{t}^2}{(\alpha_r+r)(r+\tilde{t})(\tilde{t}+\alpha_t)(\alpha_t+\alpha_r)} + \frac{(\alpha_r+r+\tilde{w}+\alpha_w)\tilde{w}^2}{(\alpha_r+r)(r+\tilde{w})(\tilde{w}+\alpha_w)(\alpha_w+\alpha_r)} \\
&+ \frac{(\alpha_r+\alpha_w+1+\alpha_t)}{(\alpha_r+\alpha_w)(\alpha_w+1)(\alpha_t+1)(\alpha_t+\alpha_r)} \Bigg] - \frac{1}{20}\Bigg[\frac{1}{r+\alpha_r} + \frac{1}{\alpha_w+\alpha_r} + \frac{1}{\alpha_t+\alpha_r} \Bigg] \Bigg)
\end{aligned}$$

For convenience reasons, the normalized width $\tilde{w} = w_m/l_m$ and thickness $\tilde{t} = t_m/l_m$ as well as the following abbreviations have been introduced,

$r = \sqrt{\tilde{w}^2 + \tilde{t}^2} \qquad \alpha_w = \sqrt{\tilde{w}^2+1}, \qquad \alpha_t = \sqrt{\tilde{t}^2+1}, \qquad \alpha_r = \sqrt{\tilde{w}^2+\tilde{t}^2+1}.$

Table A.1: Expression of the self-inductance of a rectangular brick from [72]. The dimensions are chosen according to the conductor m presented in Figure A.1.

Table A.1. This form being suitable for numerical implementations has been published by Ruehli in 1972 [72].

Mutual Inductance of two Parallel Rectangular Conductors In the case of the mutual inductance between two parallel conductors m and n according to the setup of Figure A.1, a solution has been presented in [131] on the basis of a weighted sum of 64 self-inductances of virtual conductors according to

$$L_{mn} = \frac{1}{w_m t_m w_n t_n} \frac{1}{8} \sum_{i_0 i_1 j_0 j_1 k_0 k_1 = 0}^{1} (-1)^{i_0+i_1+j_0+j_1+k_0+k_1+1} A^2_{P_{i_0 j_0 k_0} Q_{i_1 j_1 k_1}} L_{P_{i_0 j_0 k_0} Q_{i_1 j_1 k_1}}. \tag{A.2}$$

The necessary geometrical parameters are specified in Figure A.2. The self-inductance terms $L_{P_{i_0 j_0 k_0} Q_{i_1 j_1 k_1}}$ of the virtual conductors can be computed by the expression of Table A.1. Depending on the geometrical setup, some of the 64 virtual conductors may

Appendix A Partial Network Elements

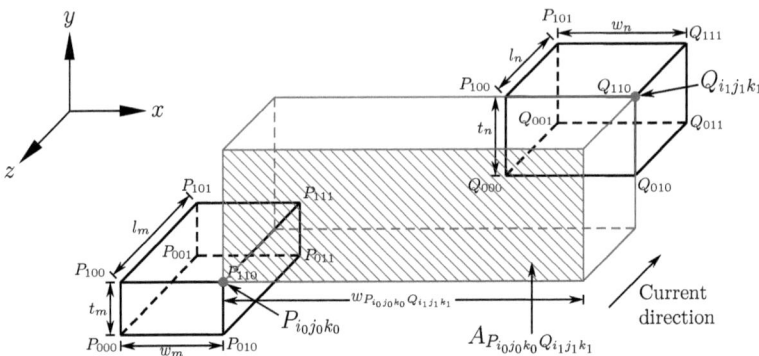

Figure A.2: Visualization of the mutual inductance concept between two parallel rectangular conductors (black corners) as a weighted sum of 64 self-inductances of virtual conductors (highlighted corners). According to (A.2), the virtual conductors are determined by choosing the two points $P_{i_0 j_0 k_0}$ and $Q_{i_1 j_1 k_1}$ in such a way that the first point is scanned over all 8 corner points of the first conductor while the latter is scanned over the corner points of the second conductor, respectively.

have zero cross section or zero length and consequently do not contribute to the overall inductance. Although the effort to compute the mutual inductance by (A.2) is up to 64 times higher than the self-inductance expression of Table A.1, it is preferred over the equation presented by Hoer and Love [130]. The reason is the fact that the evaluation is numerically more robust, especially for high aspect ratios of the conductor dimensions.

Filamentary Solution In case of long and thin conductors which are positioned in an arbitrary orientation, the double volume integral of (A.1) can be reduced to a double line integral. In this case, the conductors are regarded as filaments with the dimensions and parameters as presented in Figure A.3. In order to simplify the expressions, for each of the two filaments, a plane is introduced in such a way as to intersect with the plane of the other filament in a right angle. This allows for computing the mutual inductance between two filaments m and n as [54]

$$L_{mn} = \frac{\mu_0 \cos \varepsilon}{2\pi} \left[(\mu + l_n) \operatorname{artanh}\left(\frac{l_m}{R_1 + R_2}\right) + (\nu + l_m) \operatorname{artanh}\left(\frac{l_n}{R_1 + R_4}\right) - \mu \operatorname{artanh}\left(\frac{l_m}{R_3 + R_4}\right) - \nu \operatorname{artanh}\left(\frac{l_n}{R_2 + R_3}\right) - \frac{\Omega d}{2 \sin \varepsilon} \right], \quad \text{(A.3a)}$$

A.1 Partial Inductances

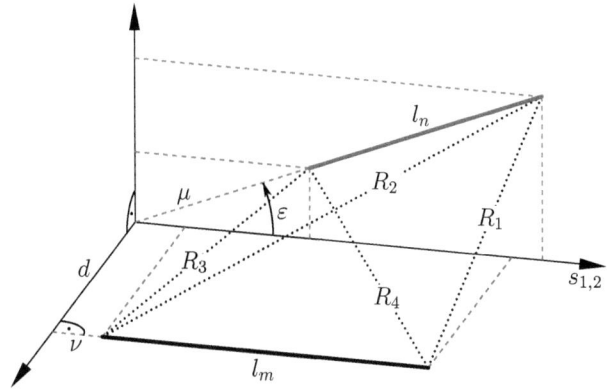

Figure A.3: Geometry and dimensions of two arbitrarily positioned filaments needed for the mutual inductance computation.

in which the introduced quantity Ω is given by

$$\Omega = \arctan\left(\frac{d^2 \cos\varepsilon + (\mu + l_n)(\nu + l_m)\sin^2\varepsilon}{d R_1 \sin\varepsilon}\right) + \arctan\left(\frac{d^2 \cos\varepsilon + \mu\nu\sin^2\varepsilon}{d R_3 \sin\varepsilon}\right) -$$
$$- \arctan\left(\frac{d^2 \cos\varepsilon + (\mu + l_n)\nu\sin^2\varepsilon}{d R_2 \sin\varepsilon}\right) - \arctan\left(\frac{d^2 \cos\varepsilon + \mu(\nu + l_m)\sin^2\varepsilon}{d R_4 \sin\varepsilon}\right). \quad \text{(A.3b)}$$

It should be noted that the solution of the above equation has been presented by earlier authors, e.g. [133]. Although (A.3) is analytically exact, numerical instabilities can occur for touching or parallel filaments. In these cases, specialized solutions as presented in [54] can be applied.

Non-orthogonal Elements Until now, the volume current cells have only been regarded for parallel arrangements of brick-shaped conductors. In order to provide solutions for more complex geometries which release the parallel precondition as well as the rectangular shape, the PEEC mesh has been extended to non-orthogonal elements in [81]. This is achieved by using local coordinates to represent (A.1). Due to the more complex structure of the obtained equation, the solution is found by numerical integration routines. A recent publication [82] combines the analytical filament approach (A.3) with the non-orthogonal volume cells, thus reducing the effort to numerically evaluate the occurring integrals.

Appendix A Partial Network Elements

A.2 Derivatives of the Partial Inductances of Rectangular Bars

In this section, the derivatives of the above closed-form partial inductance expressions w.r.t. the geometrical parameters are focused on. The results are necessitated for the adjoint sensitivity analysis in order to obtain the exact partial network derivatives [115].

Derivatives of the Self-Inductance of a Rectangular Conductor The differentiation w.r.t. the geometrical parameters is discussed for the case of conductors with rectangular cross section for which exact closed-form results have been presented in section A.1. The entire self-inductance expression of Table A.1 depending on the length l_m, the width w_m and the thickness t_m can be written as a function f_L of the two normalized variables $\tilde{w} = w_m/l_m$ and $\tilde{t} = t_m/l_m$ only, thus

$$L_{mm}(l_m, w_m, t_m) = l_m \, f_L(\tilde{w}(l_m, w_m), \tilde{t}(l_m, t_m)). \tag{A.4}$$

This allows the computation of the derivatives by applying the product and chain rules of differentiation as

$$\frac{\partial L_{mm}}{\partial l_m} = f_L - \frac{\partial f_L}{\partial \tilde{w}}\tilde{w} - \frac{\partial f_L}{\partial \tilde{t}}\tilde{t}, \quad \frac{\partial L_{mm}}{\partial w_m} = \frac{\partial f_L}{\partial \tilde{w}}, \quad \frac{\partial L_{mm}}{\partial t_m} = \frac{\partial f_L}{\partial \tilde{t}}, \tag{A.5}$$

assumed that $\partial f_L/\partial \tilde{w}$ and $\partial f_L/\partial \tilde{t}$ are known. Since the expression of Table A.1 is symmetrical w.r.t. $\tilde{w}$ and $\tilde{t}$, the differentiation of f_L must be carried out w.r.t. one of the parameters $\tilde{w}$ and $\tilde{t}$ only. In order to determine the remaining part, both arguments have to be interchanged. The differentiation of the expression of Table A.1 w.r.t. $\tilde{w}$ or $\tilde{t}$ to obtain $\partial f_L/\partial \tilde{w}$ and $\partial f_L/\partial \tilde{t}$ can be evaluated in closed-form as presented in the equation of Table A.2 which is lengthy but exact. Case studies have shown that the computational cost is about twice as much compared to the computation of the original self-inductance expression of Table A.1.

Derivatives of the Mutual Inductance of two Parallel Rectangular Conductors When differentiating (A.2) w.r.t. any shape parameter, different solutions are obtained depending on how many of the nine parameters – w_m, w_n, t_m, t_n, l_m, l_n, and three parameters for the relative shift – are perturbed when varying the shape parameter of the overall system geometry.

In the following, the results will be examined for the single conductor example of section 6.1.5 in which the overall conductor width w is varied and all sub-elements are perturbed uniformly. In this case, the solution for the derivatives of the partial mutual inductances of (A.2) w.r.t. the width of the conductor can be written in an explicit

A.3 Partial Inductances in 2D

$$\begin{aligned}\frac{\partial f_L}{\partial \tilde{w}} &= \frac{\tilde{w}}{20}\left(\frac{1}{\alpha_r(\alpha_r+\alpha_t)^2}+\frac{1}{\alpha_r^2\alpha_w+\alpha_r\alpha_w^2}+\frac{1}{\alpha_r^2 r+\alpha_r r^2}\right)+\frac{1}{6}\left(\frac{\tilde{t}^2\left(-\alpha_r^2+\tilde{w}^2\right)}{\alpha_r\tilde{w}\left(\alpha_r^2\tilde{t}^2+\tilde{w}^2\right)}+\frac{\alpha_r^2\tilde{w}+\tilde{w}^3}{\alpha_r\tilde{t}^2+\alpha_r^3\tilde{w}^2}\right.\\ &\quad -\frac{\alpha_r^2-\tilde{w}^2}{\alpha_r^3\tilde{w}+\alpha_r\tilde{t}^2\tilde{w}^3}\right)-\frac{-\alpha_w^2 r^2+\alpha_w^2\tilde{w}^2+r^2\tilde{w}^2}{4\alpha_r\alpha_w^2 r^2\tilde{w}}+\frac{\tilde{w}}{60}\left[-\frac{1}{\alpha_r\left(1+\alpha_t\right)\left(\alpha_r+\alpha_t\right)\alpha_w\left(1+\alpha_w\right)}\right.\\ &\quad +\frac{1+\alpha_r+\alpha_t+\alpha_w}{(1+\alpha_t)(\alpha_r+\alpha_t)\alpha_w(1+\alpha_w)^2(\alpha_r+\alpha_w)}+\frac{1+\alpha_r+\alpha_t+\alpha_w}{\alpha_r(1+\alpha_t)(\alpha_r+\alpha_t)^2(1+\alpha_w)(\alpha_r+\alpha_w)}+\frac{(\alpha_w+r)\tilde{w}}{\alpha_r\alpha_w(\alpha_r+\alpha_w)r(\alpha_r+r)}\\ &\quad +\frac{1+\alpha_r+\alpha_t+\alpha_w}{(\alpha_r+\alpha_t+\alpha_r\alpha_t+\alpha_t^2)(\alpha_t^2+\alpha_r\alpha_w)(\alpha_w+\alpha_w^2)}-\frac{\alpha_w r+\alpha_r(\alpha_w+r)}{\alpha_r\alpha_w(\alpha_r+\alpha_w)r(\alpha_r+r)}+\frac{-\alpha_w r+\alpha_r(3\alpha_w+r)}{\alpha_r\alpha_w(\alpha_r+r)(-\alpha_w+r)(r+\tilde{w})}\\ &\quad +\frac{\tilde{t}^2(\alpha_r+\alpha_t+r+\tilde{t})}{(\alpha_r+\alpha_t)r(\alpha_r+r)(\alpha_t+\tilde{t})(r+\tilde{t})^2}+\frac{\tilde{t}^2(\alpha_r+\alpha_t+r+\tilde{t})}{\alpha_r(\alpha_r+\alpha_t)^2(\alpha_r+r)(\alpha_t+\tilde{t})(r+\tilde{t})}+\frac{\tilde{t}^2(\alpha_r+\alpha_t+r+\tilde{t})}{(\alpha_r+\alpha_t)r(\alpha_r+r)(\alpha_t+\tilde{t})(r+\tilde{t})}\\ &\quad +\frac{2}{(\alpha_r+\alpha_w)r(\alpha_r+r)}+\frac{2}{\alpha_w(\alpha_r+\alpha_w)(\alpha_r+r)}+\frac{-\alpha_w r+\alpha_r(\alpha_w+3r)}{\alpha_r(\alpha_r+\alpha_w)(\alpha_w-r)r(\alpha_w+\tilde{w})}-\frac{\tilde{t}^2}{\alpha_r(\alpha_r+\alpha_t)r(\alpha_t+\tilde{t})(r+\tilde{t})}\right]\\ &\quad +\frac{1}{24}\left(\frac{\tilde{t}^2(2\alpha_r(\alpha_r+\alpha_t)r^2-(\alpha_r(\alpha_r+\alpha_t)+r^2)\tilde{w}^2)}{\alpha_r(\alpha_r+\alpha_t)r^2\tilde{w}\sqrt{(\alpha_r+\alpha_t)^2r^2\tilde{t}^2+\tilde{w}^4}}-\frac{\tilde{w}(\alpha_w^2\tilde{w}^2+\alpha_r r(\alpha_w^2+\tilde{w}^2))}{\alpha_r\alpha_w^2 r\sqrt{\tilde{t}^2+\alpha_w^2(\alpha_r+r)^2\tilde{w}^2}}+\frac{\tilde{t}^2(\alpha_r r-\tilde{w}^2)/(\alpha_r r)}{\tilde{w}\sqrt{\alpha_t^2(\alpha_r+r)^2\tilde{t}^2+\tilde{w}^2}}\right.\\ &\quad +\frac{2\alpha_r(\alpha_r+\alpha_t)\alpha_w^2-(\alpha_r(\alpha_r+\alpha_t)+\alpha_w^2)\tilde{w}^2}{\alpha_r(\alpha_r+\alpha_t)\alpha_w^2\tilde{w}\sqrt{(\alpha_r+\alpha_t)^2\alpha_w^2+\tilde{t}^2\tilde{w}^4}}-\frac{\tilde{w}(r^2\tilde{w}^2+\alpha_r\alpha_w(r^2+\tilde{w}^2))}{\alpha_r\alpha_w r^2\sqrt{\tilde{t}^4+(\alpha_r+\alpha_w)^2r^2\tilde{w}^2}}+\frac{(\alpha_r\alpha_w-\tilde{w}^2)/(\alpha_r\alpha_w)}{\tilde{w}\sqrt{\alpha_t^2(\alpha_r+\alpha_w)^2+\tilde{t}^4\tilde{w}^2}}\\ &\quad -\frac{\tilde{t}^2}{\tilde{w}^2}\operatorname{arsinh}\left[\frac{\tilde{w}}{\alpha_t(\alpha_r+r)\tilde{t}}\right]+\frac{2\tilde{w}}{\tilde{t}^2}\operatorname{arsinh}\left[\frac{\tilde{t}^2}{(\alpha_r+\alpha_w)r\tilde{w}}\right]-\frac{2}{t\tilde{w}^3}\operatorname{arsinh}\left[\frac{\tilde{t}\tilde{w}^2}{(\alpha_r+\alpha_t)\alpha_w}\right]-\frac{6}{\tilde{w}^2}\operatorname{arsinh}\left[\frac{\tilde{w}}{\alpha_t}\right]\\ &\quad +\frac{2\tilde{w}}{\tilde{t}}\operatorname{arsinh}\left[\frac{\tilde{t}}{\alpha_w(\alpha_r+r)\tilde{w}}\right]-\frac{1}{\tilde{t}^2\tilde{w}^2}\operatorname{arsinh}\left[\frac{\tilde{t}^2\tilde{w}}{\alpha_t(\alpha_r+\alpha_w)}\right]-\frac{2\tilde{t}^2}{\tilde{w}^3}\operatorname{arsinh}\left[\frac{\tilde{w}^2}{(\alpha_r+\alpha_t)r\tilde{t}}\right]\right)\\ &\quad +\frac{1}{6}\left[-\frac{1}{t}\operatorname{artanh}\left(\frac{\tilde{t}}{\alpha_r\tilde{w}}\right)+\frac{\tilde{t}}{\tilde{w}^2}\operatorname{artanh}\left(\frac{\tilde{w}}{\alpha_r\tilde{t}}\right)+\frac{1}{t\tilde{w}^2}\operatorname{artanh}\left(\frac{\tilde{t}\tilde{w}}{\alpha_r}\right)\right]\end{aligned}$$

Table A.2: Expression of the derivative of the self inductance of a rectangular conductor to be used with (A.5). The symbols are chosen according to Table A.1.

form. For the computation of the derivatives of the partial inductances $\partial L_{mn}/\partial w$, (A.2) is differentiated w.r.t. the total conductor width w, leading to

$$\frac{\partial L_{mn}}{\partial w} = \frac{1}{w_m t_m w_n t_n}\frac{1}{8}\sum_{i_0 i_1 j_0 j_1 k_0 k_1 = 0}^{1}\left[(-1)^{i_0+i_1+j_0+j_1+k_0+k_1+1}\right]\cdot$$
$$\cdot A^2_{P_{i_0 j_0 k_0} Q_{i_1 j_1 k_1}} \frac{w_{P_{i_0 j_0 k_0} Q_{i_1 j_1 k_1}}}{w}\frac{\partial L_{P_{i_0 j_0 k_0} Q_{i_1 j_1 k_1}}}{\partial w_{P_{i_0 j_0 k_0} Q_{i_1 j_1 k_1}}}. \quad (A.6)$$

As a closed-form expression exists for $\partial L_{mm}/\partial w_m$ of (A.5), the derivatives in (A.6) can be computed analytically.

Appendix A Partial Network Elements

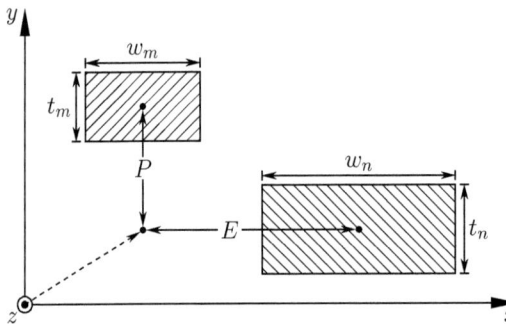

Figure A.4: Cross sections of two rectangular conductors needed for the self and mutual inductance computations in the 2D case.

A.3 Partial Inductances in 2D

In the 2D case, the partial per-unit-length inductances can be computed analytically in case of conductors with rectangular cross section as specified in Figure A.4. The solution to (4.29) can be expressed in the following form [91]

$$L'_{mn} = \frac{\mu_0}{2\pi}\left[\frac{25}{12} - \frac{1}{2w_m w_n t_m t_n}\sum_{i=1}^{4}\sum_{j=1}^{4}(-1)^{i+j}f(q_i, r_j)\right], \quad (A.7a)$$

with the following quantities $f(q_i, r_j)$, $\mathbf{q} = [q_1, q_2, q_3, q_4]$ and $\mathbf{r} = [r_1, r_2, r_3, r_4]$ as

$$f(q_i, r_j) = \left(\frac{q_i^2 r_j^2}{4} - \frac{q_i^4}{24} - \frac{r_j^4}{24}\right)\ln(q_i^2 + r_j^2) + \frac{q_i^3 r_j}{3}\arctan\left(\frac{r_j}{q_i}\right) + \frac{q_i r_j^3}{3}\arctan\left(\frac{q_i}{r_j}\right) \quad (A.7b)$$

$$\mathbf{q} = \left[E - \frac{w_m}{2} - \frac{w_n}{2},\, E + \frac{w_m}{2} - \frac{w_n}{2},\, E + \frac{w_m}{2} + \frac{w_n}{2},\, E - \frac{w_m}{2} + \frac{w_n}{2}\right] \quad (A.7c)$$

$$\mathbf{r} = \left[P - \frac{t_m}{2} - \frac{t_n}{2},\, P + \frac{t_m}{2} - \frac{t_n}{2},\, P + \frac{t_m}{2} + \frac{t_n}{2},\, P - \frac{t_m}{2} + \frac{t_n}{2}\right]. \quad (A.7d)$$

A.4 Partial Coefficients of Potential

According to the partial inductances from above, this section concentrates on the evaluation of the partial coefficients of potential for surface charges in the non-retarded case (4.14a). The expression is repeated as

$$P_{iq} = \frac{1}{4\pi\varepsilon_0 S_i S_q}\int_{S_i}\int_{S'_q}\frac{1}{|\vec{r}-\vec{r}\,'|}\,\mathrm{d}A'\,\mathrm{d}A. \quad (A.8)$$

This equation tends to be easier to evaluate than (A.1) since a double surface integral has to be evaluated instead of a double volume integral.

A.4 Partial Coefficients of Potential

In the past, various publications have concentrated on the evaluation of (A.8) for different geometries. In case of rectangular elementary patches such as presented in Figure A.5, results will be discussed in the following paragraph. In this case, it is stated in [134] that the solution can be regarded as a special case of the inductance calculations of (A.1). Besides rectangular patches, there have also been investigations on solving (A.8) for triangular patches as these may approximate various geometries more flexible, e. g. [135, 136].

If the collocation method with Dirac-delta shaped testing functions is used in the PEEC method, an alternative solution to (A.8) with a single surface integral is obtained. Since the elements are easier to compute, some papers [46, 137, 138] focus on this case. However, the coefficients of potentials are no longer symmetrically. Moreover, a finer mesh is generally required to achieve similar accuracies compared to the Galerkin approach resulting in (A.8).

Coefficients of Potential for Rectangular Patches For two parallel oriented rectangular patches according to Figure A.5a, the coefficient of potential of (A.8) can solved as [73]

$$P_{iq} = \frac{1}{4\pi\varepsilon_0 f_a f_b s_a s_b} \sum_{k=1}^{4} \sum_{m=1}^{4} (-1)^{k+m} \left[\frac{b_m^2 - c_{iq}^2}{2} a_k \ln(a_k + \varrho) - \frac{1}{6}(b_m^2 - 2c_{iq}^2 + a_k^2)\varrho \right.$$
$$\left. + \frac{a_k^2 - c_{iq}^2}{2} b_m \ln(b_m + \varrho) - b_m\, c_{iq}\, a_k \arctan\left(\frac{a_k b_m}{\varrho c_{iq}}\right) \right], \quad \text{(A.9a)}$$

in which the following abbreviations have been introduced

$$\varrho = \sqrt{a_k^2 + b_m^2 + c_{iq}^2}, \tag{A.9b}$$

$$\mathbf{a} = \left[a_{iq} - \frac{f_a}{2} - \frac{s_a}{2},\ a_{iq} + \frac{f_a}{2} - \frac{s_a}{2},\ a_{iq} + \frac{f_a}{2} + \frac{s_a}{2},\ a_{iq} - \frac{f_a}{2} + \frac{s_a}{2} \right], \tag{A.9c}$$

$$\mathbf{b} = \left[b_{iq} - \frac{f_b}{2} - \frac{s_b}{2},\ b_{iq} + \frac{f_b}{2} - \frac{s_b}{2},\ b_{iq} + \frac{f_b}{2} + \frac{s_b}{2},\ b_{iq} - \frac{f_b}{2} + \frac{s_b}{2} \right]. \tag{A.9d}$$

As before, special attention has to be paid to the case when the parallel panels touch each other because both, the ln-function as well as the associated factor approach infinity and zero, respectively. For the special case in which both patches are located on the same plane, i. e. $c_{iq} = 0$, an alternative formulation is presented in [139]. Moreover, the self-coefficient of potential P_{ii} can also be computed by (A.9).

Appendix A Partial Network Elements

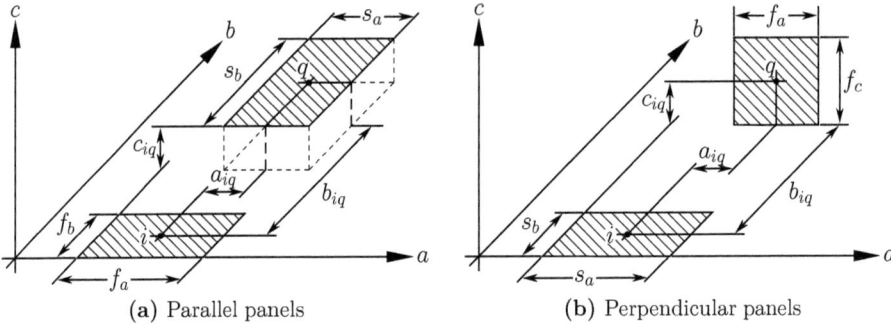

Figure A.5: Geometry of the parallel and perpendicular panel setups for the computation of the partial coefficients of potential according to [73].

If both patches are oriented perpendicular to each other (s. Figure A.5b), the following solution is obtained [73]

$$P_{iq} = \frac{1}{4\pi\varepsilon_0 f_a f_c s_a s_b} \sum_{k=1}^{4}\sum_{m=1}^{2}\sum_{l=1}^{2}(-1)^{k+m+l+1}\left[\left(\frac{a_k^2}{2}-\frac{c_l^2}{6}\right)c_l \ln(b_m+\varrho) - \frac{b_m c_l}{3}\varrho + \right.$$
$$\left. + \left(\frac{a_k^2}{2}-\frac{b_m^2}{6}\right)b_m \ln(c_l+\varrho) + a_k b_m c_l \ln(a_k+\varrho) - \frac{a_k^3}{6}\arctan\left(\frac{b_m c_l}{a_k \varrho}\right) - \right.$$
$$\left. - \frac{b_m^2 a_k}{2}\arctan\left(\frac{a_k c_l}{b_m \varrho}\right) - \frac{a_k c_l^2}{2}\arctan\left(\frac{a_k b_m}{c_l \varrho}\right)\right], \quad \text{(A.10a)}$$

in which ϱ and $\mathbf{a}$ are equal to (A.9b) and (A.9c). Additionally, the following quantities have been introduced

$$\mathbf{b} = \left[b_{iq}+\frac{s_b}{2},\ b_{iq}-\frac{s_b}{2}\right], \qquad \mathbf{c} = \left[c_{iq}+\frac{f_c}{2},\ c_{iq}-\frac{f_c}{2}\right]. \quad \text{(A.10b)}$$

A.5 Static Green's Function of a Two-layer Substrate

In the last section of this appendix, the coefficient-of-potential definitions for parallel rectangular patches of (A.9) are extended to the case in which the conductors are located on a dielectric material. The dielectric layer is visualized in Figure A.6 and is assumed to be extended towards infinity in both x- and y-directions. In addition, the substrate thickness h and the conductor widths must be much larger than the thickness t of the conductors, thus allowing a treatment of the conductors as patches and additionally a consideration of the z-values $z=0$ and $z=h$ only.

A.5 Static Green's Function of a Two-layer Substrate

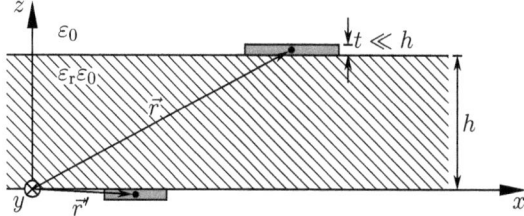

Figure A.6: Cross section of a two-conductor arrangement on a dielectric substrate with two layers. The conductor thickness is assumed to be much smaller than the substrate thickness.

In this setup, it is convenient to rewrite the quasi-stationary Green's function of free space (2.32) in cartesian coordinates according to

$$\hat{G}(\vec{r},\vec{r}')=\frac{1}{|\vec{r}-\vec{r}'|}=\frac{1}{\sqrt{(x-x')^2+(y-y')^2+(z-z')^2}}=\frac{1}{\sqrt{\varrho^2+\Delta z^2}}, \quad (A.11a)$$

with the abbreviations

$$\varrho=\sqrt{(x-x')^2+(y-y')^2}, \qquad \Delta z = z-z'. \quad (A.11b)$$

The above Green's function of free space can be adapted to the setup of Figure A.6 by applying the method of images. For the case that both, source and observation points are located in the $z=0$ plane, the following Green's function is obtained [46]

$$\hat{G}(\vec{r},\vec{r}')_{\text{SameLayers}}=\frac{1}{4\pi}\left[\frac{1-\tilde{\varepsilon}_r}{\varrho}+(1-\tilde{\varepsilon}_r^2)\sum_{k=1}^{\infty}\frac{\tilde{\varepsilon}_r^{2k-1}}{\sqrt{\varrho^2+(2kh)^2}}\right]. \quad (A.12a)$$

If both, source- and observation points are positioned on the opposite layers with $z'=0$ and $z=h$, the Green's function becomes [46]

$$\hat{G}(\vec{r},\vec{r}')_{\text{OppLayers}}=\frac{1-\tilde{\varepsilon}_r^2}{4\pi}\sum_{k=1}^{\infty}\frac{\tilde{\varepsilon}_r^{2(k-1)}}{\sqrt{\varrho^2+(h(2k-1))^2}}. \quad (A.12b)$$

In the above equations (A.12), the following abbreviation has been introduced

$$\tilde{\varepsilon}_r=\frac{\varepsilon_r-1}{\varepsilon_r+1}. \quad (A.12c)$$

When comparing the Green's function of free space (A.11) with the adapted Green's functions of (A.12), the free space solution of the coefficients of potential (A.9) can be transferred to the two layer arrangement. In particular, this is achieved by interchanging the integration of (A.8) and the summation of (A.12) and substituting the c_{iq} displayed in Figure A.5a by $2kh$ or $h(2k-1)$ of (A.12a) and (A.12b), respectively. Thus, the

167

Appendix A Partial Network Elements

numerical effort for computing the coefficients of potential for the dielectric substrate increases due to the fact that (A.9) has to be computed repeatedly. In [73], some hints are presented about the truncation criterion of the infinite series of (A.12).

It should be mentioned that the above concept can easily be extended to dielectric losses included in the relative permittivity. This is obtained by replacing ε_r by $\underline{\varepsilon}_r$ as stated in (2.12).

Appendix B
DC Analysis of a Rectangular Conductor Bend

In this appendix, the expressions (6.5) of page 124 allowing the computation of the DC resistance of a rectangular conductor bend with two different widths w_x and w_y are derived by means of the Schwarz-Christoffel mapping technique. The conformal mapping technique can be used to solve problems based on the Laplace equation since it is invariant to this kind of transformation.

The Schwarz-Christoffel mapping is a conformal transformation in which the upper half plane is transformed onto the interior of a polygon. In particular, the transformation from the complex w-plane to the z-plane is given by the following expression

$$z(w) = A \int \prod_{\nu=1}^{n} (w - u_\nu)^{-\frac{\alpha_\nu}{\pi}} \, dw + B, \tag{B.1}$$

in which A and B are two complex constants,[1] u_ν the real parts in the w-plane belonging to the n corners of the polygon and α_ν the rotation angles in the z-plane.

In Figure B.1, the different complex planes which are needed for the particular mapping of the straight conductor onto the rectangular conductor bend are visualized. The simple mesh of the ζ-plane corresponds to equipotential and electric field lines of a straight conductor for which the relations such as the current distribution or the resistance are well known. With the help of the intermediate transformation to the w-plane, the simple mesh of the ζ-plane is transformed onto the mesh in the z-plane which reflects the behavior of the desired conductor bend geometry.

Transformation from w to z Since the Schwarz-Christoffel transformation is based on the mapping from the upper half plane of the w-plane onto the interior of a polygon, the intermediate w-plane needs to be set up. For applying the above general expression (B.1) to the specific case, the u_ν values as well as the α_ν angles have to be determined.

---
[1] In contrast to previous chapters, complex values are not underlined in this derivation.

Appendix B DC Analysis of a Rectangular Conductor Bend

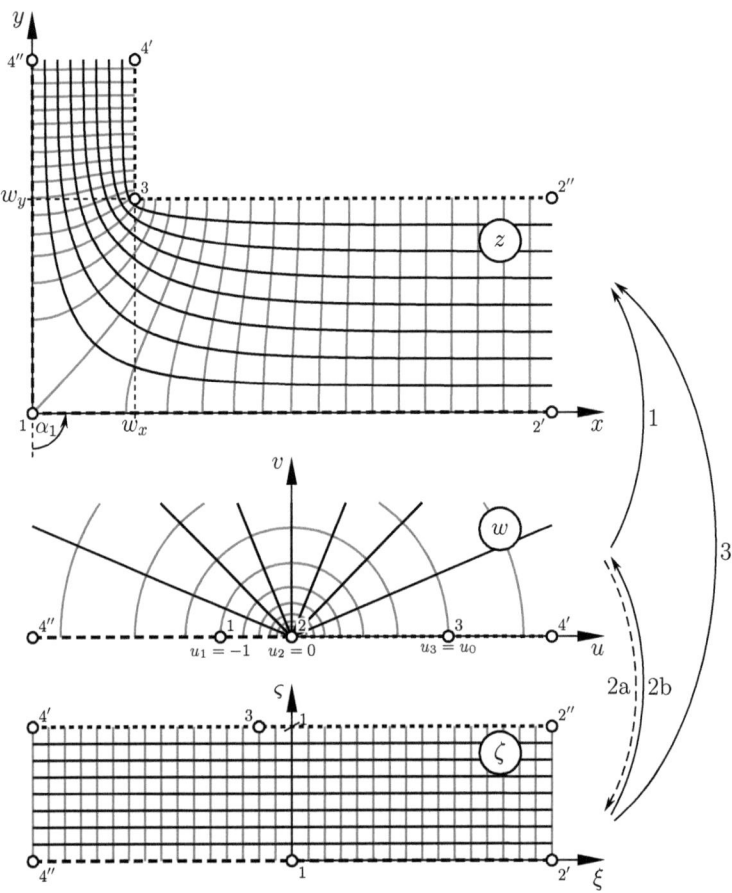

Figure B.1: Conformal mapping of the corner via Schwarz-Christoffel transformation. Two intermediate steps are required to transform the simple mesh from the ζ-plane onto the desired z-plane. The transformation is motivated by the fact that the resistance is trivial to compute in the ζ-plane.

Because two of the u_ν values can be chosen arbitrarily, the points $u_1 = -1$ and $u_2 = 0$ are fixed (s. Figure B.1). The point $u_3 = u_0$ is unknown at the beginning since it must include the information about the widths w_x and w_y.

As can be verified by the z-plane displayed in Figure B.1, the outer rotation angles in the z-plane are $\alpha_1 = \pi/2$ at point 1, $\alpha_2 = \pi$ at point[2] 2 and $\alpha_3 = -\pi/2$ at point 3. This allows for concretizing (B.1) to the following form

$$z(w) = A \int (w+1)^{-\frac{1}{2}} (w-0)^{-1} (w-u_0)^{\frac{1}{2}} \, dw + B \tag{B.2a}$$

$$= A \int \frac{\sqrt{w-u_0}}{w\sqrt{w+1}} dw + B. \tag{B.2b}$$

The integral can be solved as follows

$$z(w) = -A\sqrt{u_0} \arctan\left(\frac{w - u_0(w+2)}{2\sqrt{u_0}\sqrt{w+1}\sqrt{w-u_0}}\right) +$$
$$+ A \ln\left(1 - u_0 + 2w + 2\sqrt{w+1}\sqrt{w-u_0}\right) + B. \tag{B.2c}$$

The remaining unknowns of (B.2c) are $u_0 \in \mathbb{R}^+$, $A \in \mathbb{C}$ and $B \in \mathbb{C}$. When evaluating the point 2 which is zero in the w-plane and infinite in the real part of the z-plane, it turns out that the constant A must be purely imaginary. This allows the calculation of the three unknowns by evaluating the points 1 and 3 as

$$z(w=-1) = A\left[-\frac{\pi}{2}\sqrt{u_0} + \ln(1+u_0) + j\pi\right] + B \stackrel{!}{=} 0, \tag{B.3a}$$

$$z(w=u_0) = A\left[\frac{\pi}{2}\sqrt{u_0} + \ln(1+u_0)\right] + B \stackrel{!}{=} w_x + jw_y, \tag{B.3b}$$

which results in

$$u_0 = \frac{w_y^2}{w_x^2}, \quad A = \frac{jw_x}{\pi}, \quad B = w_x + \frac{jw_y}{2} - \frac{jw_x}{\pi} \ln\left(1 + \frac{w_y^2}{w_x^2}\right). \tag{B.4}$$

Transformation from w to ζ In the second step, the w-plane is transformed onto the ζ-plane with the same approach as before (dashed arrow 2a in Figure B.1). Afterwards, the inverse function is built (arrow 2b in Figure B.1) since the straight conductor segment on the ζ-plane has to be mapped to the corner (arrow 3 in Figure B.1).

The second transformation is simpler compared to the first one because there is only one rotation angle at the point 2. As an additional precondition, the height in η-direction is set to one from which follows the condition $\zeta(w=1) = j$. This allows for formulating the Schwarz-Christoffel transformation (B.1) with the constants C and D according to

$$\zeta(w) = C \int (w+u_0)^{-0} (w-0)^{-1} (w-1)^0 \, dw + D = C \int \frac{1}{w} dw + D \tag{B.5a}$$

$$= C \ln(w) + D. \tag{B.5b}$$

---
[2]The point 2 at infinity is also accounted for as a 180° rotation.

Appendix B DC Analysis of a Rectangular Conductor Bend

When regarding the following relations

$$\zeta(w = -1) = 0, \qquad \zeta(w = 1) = j, \tag{B.6}$$

both constants can be determined as

$$C = -\frac{1}{\pi}, \qquad D = j. \tag{B.7}$$

These values can now be substituted into (B.5b) in order to express the transformation according to

$$\zeta(w) = -\frac{1}{\pi} \ln(w) + j, \tag{B.8}$$

or the inverse function as

$$w(\zeta) = -e^{-\pi\zeta}. \tag{B.9}$$

By substituting (B.9) into (B.2c) and regarding (B.4), the desired overall transformation (arrow 3 in Figure B.1) results in the lengthy expression

$$z(\zeta) = \frac{1}{2\pi} j \left((w_y - 2jw_x)\pi - 2w_x \ln\left[1 + \frac{w_y^2}{w_x^2}\right] + \right.$$

$$+ 2w_y \operatorname{arccot}\left[\frac{2w_x w_y e^{\pi\zeta}\sqrt{1 - e^{-\pi\zeta}}\sqrt{-\frac{w_y^2}{w_x^2} - e^{-\pi\zeta}}}{w_x^2 + w_y^2(-1 + 2e^{\pi\zeta})}\right] +$$

$$\left. + 2w_x \ln\left[1 - \frac{w_y^2}{w_x^2} - 2e^{-\pi\zeta} + 2\sqrt{1 - e^{-\pi\zeta}}\sqrt{-\frac{w_y^2}{w_x^2} - e^{-\pi\zeta}}\right] \right). \tag{B.10a}$$

In the special case $w_y = w_x = w$, the above result can be simplified to

$$z(\zeta) = \frac{w}{\pi}\left[j \arccos\left(e^{\pi\zeta}\right) + \arccos\left(e^{-\pi\zeta}\right)\right], \qquad \text{for} \qquad w = w_x = w_y. \tag{B.10b}$$

When transforming different straight lines in the ζ-plane with $\operatorname{Re}\{\zeta\}$ = const. and $\operatorname{Im}\{\zeta\}$ = const., the desired mesh presented in Figure B.2 is obtained.

Resistance of the conductor bend In order to compute the resistance of the corner from the results of the previous considerations, the behavior of the two introduced points 5 and 6 displayed in Figure B.3 is analyzed. Especially the transformation of both points from the ζ-plane onto the z-plane is of importance since the resistance of the conductor in the ζ-plane (s. Figure B.3b) can be expressed by means of ξ_0 and ξ_1 according to (2.40) as

$$R_{\text{DC}} = \frac{1}{\kappa t}\left(\frac{\xi_0}{1} - \frac{\xi_1}{1}\right) = \frac{1}{\kappa t} R''. \tag{B.11}$$

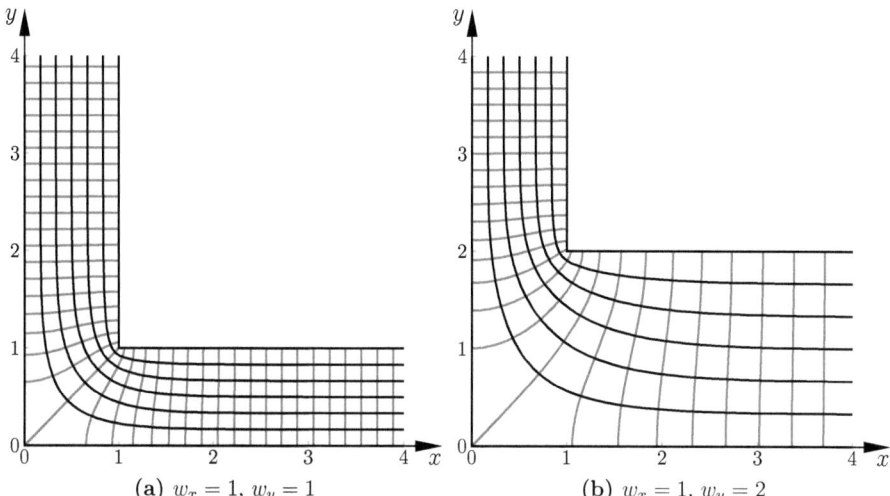

Figure B.2: Transformed mesh according to (B.10b) and (B.10a) for two exemplary settings of the conductor widths. A mesh-line separation of $1/6$ is chosen in the ζ-plane. The visualized lines can be interpreted as constant potential values and electric field lines.

In the above equation, κ and t are again the electric conductivity and the conductor thickness, respectively. Due to the fact that the 2D mesh does not depend on these two parameters, the following considerations are focused on the dimensionless geometrical resistance R'' of (B.11) only (cf. [122]). In order to obtain the transformed information of ξ_0 and ξ_1 in the z-plane of shown in Figure B.3a, the limit behavior for large values is analyzed in the following.

At first, the point 5 which is $\zeta_0 = \xi_0 + j\,0$ in Figure B.3b, is converted to the z-plane. By substituting this point into (B.10a) and building the limit value for large ξ_0, the result can be expressed in the following form[3]

$$\lim_{\xi_0 \to \infty} z(\xi_0) = w_y\,\xi_0 + w_x - \frac{2w_x}{\pi}\arctan\left(\frac{w_y}{w_x}\right) - \frac{w_y}{\pi}\ln\left[\frac{1}{4}\left(1+\frac{w_x^2}{w_y^2}\right)\right]. \quad \text{(B.12a)}$$

Due to the fact that (B.12a) is real valued only, the expression is assigned with $x_0 = z(\xi_0)$, $\xi_0 > 0$ in the following which can be verified by Figure B.3a. A further important property of (B.12a) is the fact that x_0 increases linearly with ξ_0.

---

[3] The expression (B.12a) is obtained by converting the arccot-function of (B.10a) to the ln-function and afterwards building the limit for each part separately. The limit of one specific part is difficult to handle whereas it is convenient to express an occurring $\sqrt{1+x}$ term by the first three parts of the corresponding series representation.

Appendix B DC Analysis of a Rectangular Conductor Bend

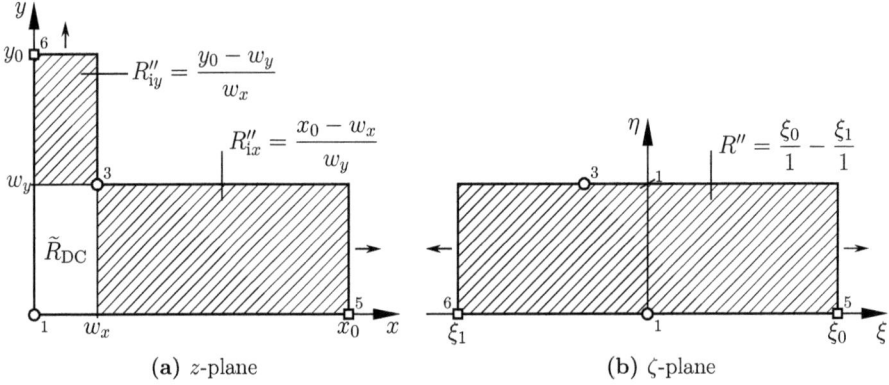

Figure B.3: Computation of the resistance via the introduced points 5 and 6. The resistances of both setups are compared by introducing the hatched regions and assigning the difference to the white area of (a) indicated by $\tilde{R}_{\text{DC}}$.

Equivalent to the limit behavior of ξ_0, the point ξ_1 which is referred to as point 6 in Figure B.3, can be expressed as

$$\lim_{\xi_1 \to -\infty} z(\xi_1) = j\left(-w_x\,\xi_1 + w_y - \frac{2w_y}{\pi}\arctan\left(\frac{w_x}{w_y}\right) - \frac{w_x}{\pi}\ln\left[\frac{1}{4}\left(1 + \frac{w_y^2}{w_x^2}\right)\right]\right), \quad \text{(B.12b)}$$

and shows an imaginary component $y_0 = \operatorname{Im}\{z(\xi_1)\}$, $\xi_1 < 0$ only which increases linearly with ξ_1.

The linear dependence of both points x_0 and y_0 with the corresponding points ξ_0 and ξ_1 for large values allows for comparing both resistances of the ζ-plane and the z-plane. In particular, the hatched areas according to Figure B.3 are considered. In there, a difference resistance $\tilde{R}_{\text{DC}}$ is introduced which can graphically be interpreted as the white area visualized in Figure B.3a. This area is indicated with $\tilde{R}_{\text{DC}}$ being the difference of both planes according to

$$\tilde{R}_{\text{DC}} = R'' - R_i'' = \left(\frac{\xi_0}{1} - \frac{\xi_1}{1}\right) - \left(\frac{x_0(\xi_0) - w_x}{w_y} + \frac{y_0(\xi_1) - w_y}{w_x}\right). \quad \text{(B.13)}$$

In the above equation, $R_i'' = R_{ix}'' + R_{iy}''$ denotes the sum of both hatched areas according to Figure B.3a and corresponds to the inner dimensions of the conductors connected at the corner.

By substituting x_0 and y_0 in (B.13) by the results of (B.12a) and (B.12b) for the limiting case $\xi_0 \to \infty$ and $\xi_1 \to -\infty$, the linear dependencies eliminate each other and

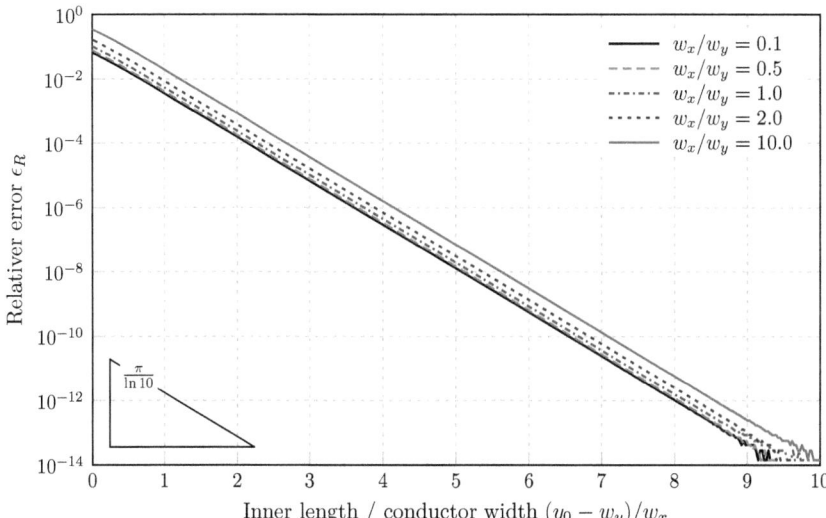

Figure B.4: Relative resistance error for increasing lengths of the straight conductors. If the inner length of the conductor is at least twice the width, the relative resistance error of the approach using (B.14) is already below 0.1 %.

the following expression is obtained

$$\tilde{R}_{\text{DC}} = \frac{w_y}{w_x} + \frac{2}{\pi} \ln\left(\frac{w_x^2 + w_y^2}{4\,w_x w_y}\right) + \frac{2}{\pi} \frac{w_x^2 - w_y^2}{w_x w_y} \arctan\left(\frac{w_y}{w_x}\right), \quad \text{(B.14a)}$$

which is a function of the geometrical parameters w_x and w_y only. If both conductors hold the same width $w_x = w_y = w$, the correction term simplifies to

$$\tilde{R}_{\text{DC}} = 1 - \frac{2\ln 2}{\pi} \approx 0.5587, \qquad \text{for} \qquad w_x = w_y, \quad \text{(B.14b)}$$

which is in accordance with [122]. The above result states that the correct resistance of the conductor bend can be computed by taking the resistances of the inner dimensions and adding the correction term $\tilde{R}_{\text{DC}}$ according to the above equations (B.14).

Since this term is exact only when $\xi_0 \to \infty$ und $\xi_1 \to -\infty$ approach infinity, the errors for different lengths of the straight conductors are analyzed. For this reason, a setup is constructed in which ξ_0 is fixed at a very large value and ξ_1 is swept from $\xi_1 = 0$ to $\xi_1 = -10$. For each length, the relative resistance error

$$\epsilon_R(\xi_1) = \frac{\left|R''(\xi_1) - \left[R_i''(\xi_1) + \tilde{R}_{\text{DC}}\right]\right|}{|R''(\xi_1)|} \quad \text{(B.15)}$$

175

Appendix B DC Analysis of a Rectangular Conductor Bend

is computed which includes the correction term $\tilde{R}_{\text{DC}}$ even for finite conductor lengths. The results for different w_x/w_y ratios are plotted in Figure B.4 as a function of the inner conductor length normalized to the conductor width. It can be seen that the resistance error decreases exponentially and shows similar behavior for widths ratios from 0.1 to 10. If the ratio of the inner conductor length by the conductor width is at least two, the relative resistance error is already below 0.1 %. For an increased ratio of five, the error decreases below 10^{-7} whereas it is numerically negligible for long and thin conductors with a length of more than 10 times the width.

Appendix C
Skin-Effect Discretization of a Rectangular Conductor

In this appendix, the cross sectional discretization of rectangular conductors for skin-effect applications is investigated. Especially, the skin factor χ which influences the number of mesh cells on the one hand and the accuracy on the other hand has to be chosen carefully. The skin factor has been introduced in Figure 4.11 and accounts for the estimated non-uniform current distribution inside the conductors by determining the ratio of the sizes of two neighboring cells.

In order to select an adequate value of the skin factor, the following two test scenarios as depicted in Figures C.1 and C.2 are set up. Two conductors with different cross sections are analyzed at three different frequencies for different discretization factors χ and different maximum sizes of the outermost segments δ_{wt}/δ. For each simulated parameter setup, the error is computed w.r.t. the reference simulation obtained by the convergence analysis presented in Figure 6.9.[1]

The results can be interpreted as follows: The four different curves in each plot of Figures C.1 and C.2 belong to four different sizes of the outermost corner elements shown in Figure 4.11. For $\delta_{wt} = \delta$, the widths of the corner elements are equal[2] to the skin depth of the actual frequency. For the three remaining curves, the size is decreased to $\delta/2$, $\delta/4$ and $\delta/8$, respectively. When regarding the left parts of the figures belonging to a skin factor near 1, the relative error is lower for smaller element size as expected. However, if the skin factor is increased, the error may grow although the outermost element size remains unchanged. The reason is the fact that by increasing the skin factors, the error is dominated by the larger elements in the interior of the conductor since the proportion of two neighboring cells is increased.

In order to find an adequate parameter setting as a tradeoff between effort and accuracy, different horizontal lines have been appended to the figures, indicating the maxi-

---

[1] The reference solutions can be estimated to have an accuracy of approximately 10^{-4}.

[2] In this study, the subdivision procedure of (4.36) is slightly adapted in order to exactly match the size of the outermost segments.

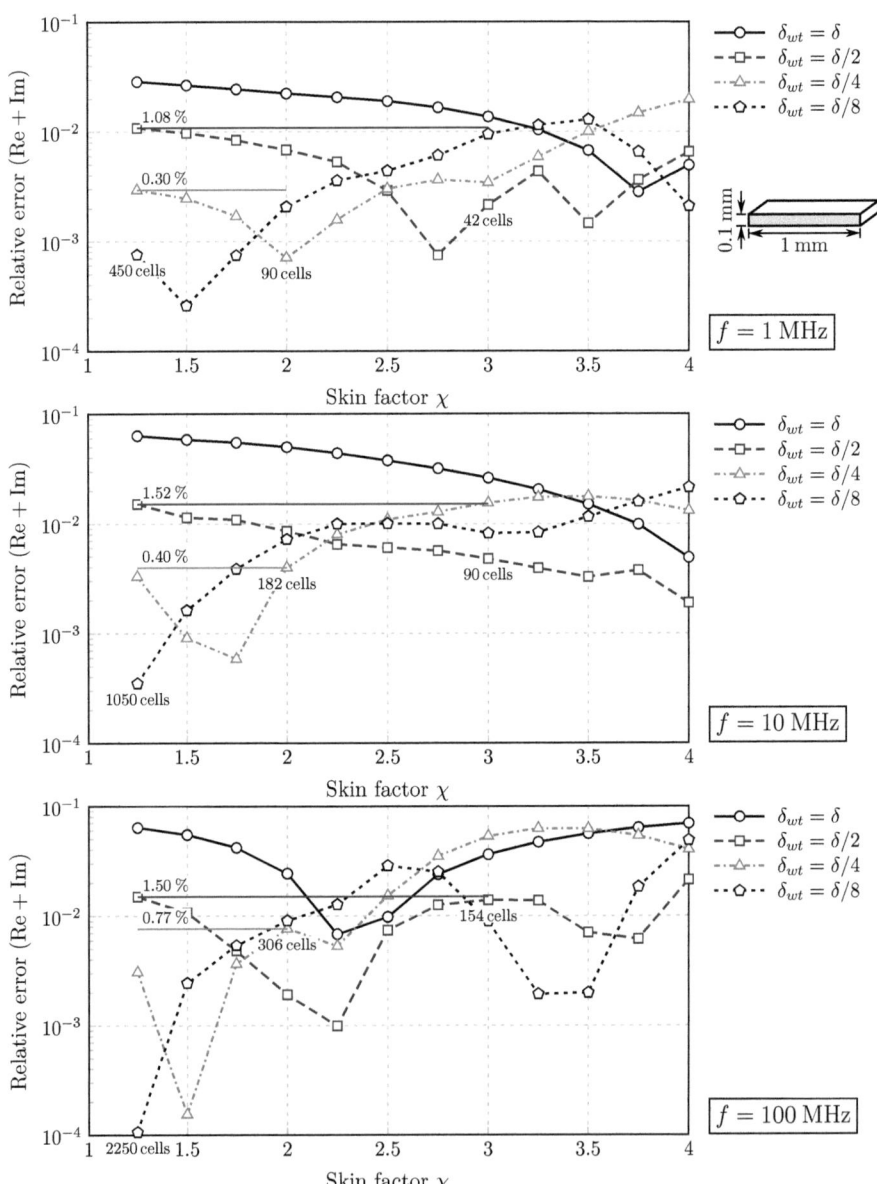

Figure C.1: Comparison of the discretization error of a conductor with $l = 50$ mm, $w = 1$ mm, $t = 0.1$ mm and $\kappa = 58\,10^6$ S/m. Results are summarized in Table C.1.

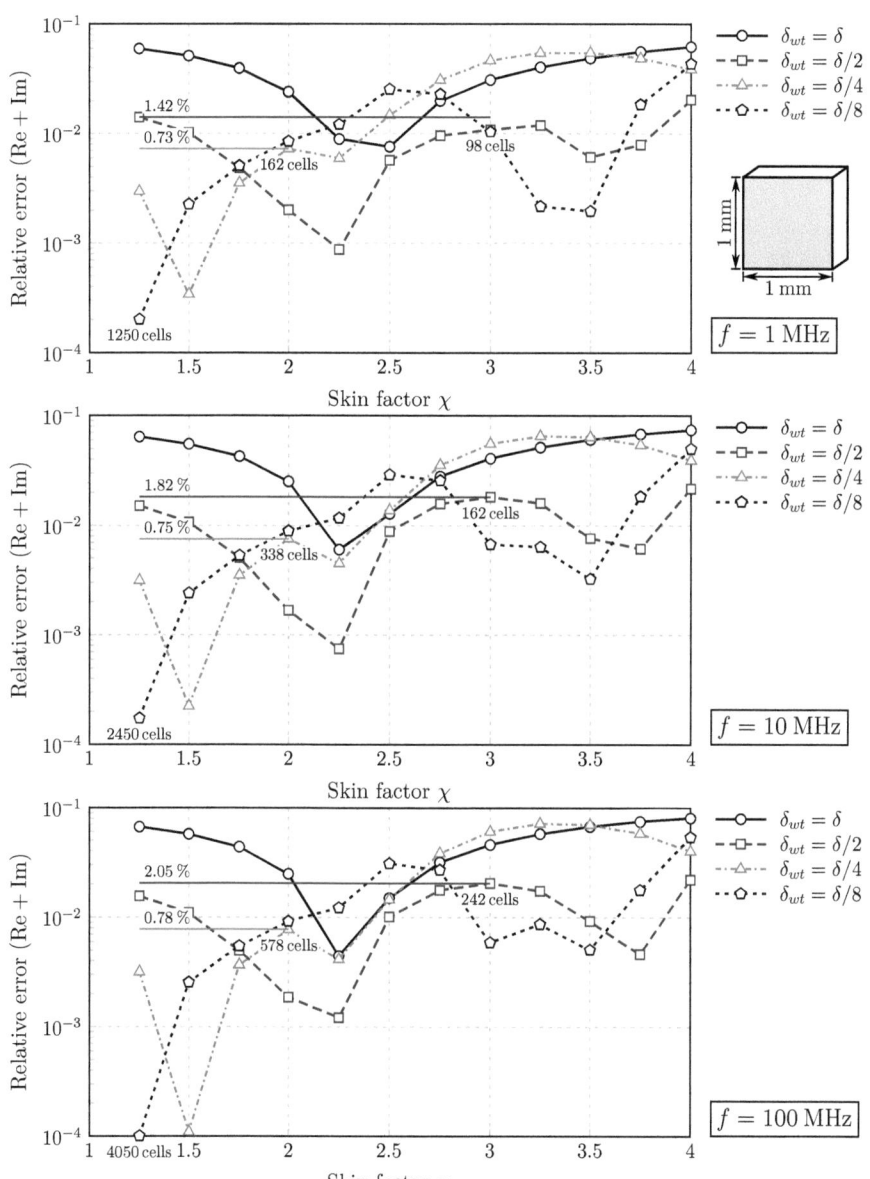

Figure C.2: Comparison of the discretization error of a conductor with $l = 50$ mm, $w = 1$ mm, $t = 1$ mm and $\kappa = 58\,10^6$ S/m. Results are summarized in Table C.1.

Appendix C Skin-Effect Discretization of a Rectangular Conductor

#	χ	δ_{wt}	Error	DoF/DoF$_{\#1}$
1	3	$\delta/2$	$\approx 2\,\%$	1
2	2	$\delta/4$	$< 1\,\%$	≈ 2
3	1.25	$\delta/8$	$< 0.1\,\%$	≈ 14

Table C.1: Summarized results of Figures C.1 and C.2.

mum error in a certain skin factor range. Starting with the straight lines belonging to $\delta_{wt} = \delta/2$, the error ranges from approximately 1 % to 2 % for skin factors of in between 1.25 and 3. If lower errors of less than 1 % are demanded, not only the outermost element size has to be decreased to $\delta_{wt} = \delta/4$, but also the maximum skin factor should not significantly exceed $\chi = 2$. When regarding the required effort for the decreased error, the necessary number of cells doubles roughly as can be seen from the selected cell numbers of Figures C.1 and C.2. The demanded effort even more increases if errors below 0.1 % are required. In this case, a skin factor of less or equal than 1.25 should be chosen while setting the outermost element size close to $\delta_{wt} = \delta/8$.

In Table C.1, the results are summarized for three different χ-δ_{wt} settings. When comparing the first and the third setting, the error can be decreased by more than one order of magnitude. At the same time, the number of unknowns has increased by a factor of about 14 which is not affordable for larger problems.

For the applications used in this work, $\delta_{wt} \leq \delta/2$ is chosen with a skin factor of $\chi = 2$. This setting provides slightly better results compared to the first case of Table C.1. However, one has to keep in mind that errors of approximately 1 % are obtained by using this technique. Due to the fact that the outermost element size is approximately half of the size of the skin depth, this discretization setting is also referred to as $\delta/2$-rule.

Acronyms and Symbols

Acronyms

1D	One Dimensional
2D	Two Dimensional
2D-PEEC	Two Dimensional – Partial Element Equivalent Circuit
3D	Three Dimensional
AC	Alternating Current
AD	Automatic Differentiation
BEM	Boundary Element Method
BLC	Boundary-Layer Concept
DC	Direct Current
DC-PEEC	Direct Current – Partial Element Equivalent Circuit
DoF	Degrees of Freedom
ECF	Equivalent Charge Formulation
EFIE	Electric Field Integral Equation
EM	Electromagnetic
EMQS	Electro-Magneto-Quasi-Static
EQS	Electro-Quasi-Static
FD	Finite Difference
FEM	Finite Element Method
FIT	Finite Integration Technique
FMM	Fast Multipole Method
HF	High Frequency
IC	Integrated Circuit
ILC	Inner-Layer Concept
IPT	Inductive Power Transfer
ISM	Industrial, Scientific and Medical
KCL	Kirchhoff's Current Law
KVL	Kirchhoff's Voltage Law
LQS	Lorenz-Quasi-Static
LQS-PEEC	Lorenz-Quasi-Static – Partial Element Equivalent Circuit
LU	Lower and Upper triangular matrices
MN	Matching Network

Acronyms

MNA	Modified Nodal Analysis
MoM	Method of Moments
MPIE	Mixed Potential Integral Equation
MQS	Magneto-Quasi-Static
MQS-PEEC	Magneto-Quasi-Static – Partial Element Equivalent Circuit
PCB	Printed Circuit Board
PEC	Perfect Electric Conductor
PEEC	Partial Element Equivalent Circuit
PSC	Printed Spiral Coil
QS	Quasi-Static
RAM	Random-Access Memory
RF	Radio Frequency
RFID	Radio Frequency Identification
RLC	Network of resistances, inductances and capacitances
RMS	Root Mean Square
rPEEC	Retarded Partial Element Equivalent Circuit
SPICE	Simulation Program with Integrated Circuit Emphasis
SRF	Self-Resonant Frequency
Transceiver	*Trans*mitter-Re*ceiver*
Transponder	*Trans*mitter-Res*ponder*
VCO	Voltage Controlled Oscillator

General Symbols and Conventions

Notation	Description
$\mathbb{C}$	Complex numbers
$\mathbb{N}$	Natural numbers
$\mathbb{R}$	Real numbers
$\mathbf{X}$	Matrix
$\mathbf{x}, \mathbf{x}^{\mathrm{T}}$	Column vector and row vector (transposed)
$\vec{x}$	Spatial vector in $\mathbb{R}^3$
$\vec{x} \cdot \vec{y}$	Dot product of vectors $\vec{x}$ and $\vec{y}$
$\vec{x} \times \vec{y}$	Cross product of vectors $\vec{x}$ and $\vec{y}$
$\underline{x}$	Complex value
$\underline{x}^*$	Complex conjugate value
$\lvert x \rvert, \lvert \underline{x} \rvert$	Absolute value of scalar and complex numbers
$\lvert \vec{x} \rvert, \lvert \underline{\vec{x}} \rvert$	Absolute value of scalar and complex valued vectors
x	Scalar value
∇	Nabla operator, $\nabla = (\partial/\partial x, \partial/\partial y, \partial/\partial z)^{\mathrm{T}}$ in cart. coordinates
$\in$	"Element of" symbol
∂	Partial derivative operator

Greek Letters

Notation	Description	Unit
α	Scaling factor of a matching network	
β	Arbitrary angle	[rad]
γ	Fill factor of a spiral coil	
Δ	Laplace operator, $\Delta \Theta = \operatorname{div}\operatorname{grad} \Theta$	
Δp_n	Step size of the design parameter p_n	
Δw	Perturbation of the conductor width w	[m]
δ	Skin depth	[m]
δ_a	Max. outermost segment size of a circular conductor	[m]
δ_t	Max. outermost segment size of a cond. in t-direction	[m]
δ_w	Max. outermost segment size of a cond. in w-direction	[m]
δ_{wt}	Geometric mean value, $\delta_{wt} = \sqrt{(\delta_w \, \delta_t)}$	[m]
$\varepsilon, \underline{\varepsilon}$	Permittivity, complex value includes losses	[F m^{-1}]
ε_0	Permittivity of free space, $\varepsilon_0 \approx 8.85 \cdot 10^{-12}$ F m^{-1}	[F m^{-1}]
$\varepsilon_{\mathrm{r}}, \underline{\varepsilon}_{\mathrm{r}}$	Relative permittivity, complex value includes losses	
η	Efficiency	

Greek Letters

Notation	Description	Unit
Θ	Auxiliary scalar field	
ϑ	Inclination angle in spherical coordinates	[rad]
κ	Electric conductivity	[S m^{-1}]
λ	Wavelength	[m]
μ	Permeability of a magnetic material	[H m^{-1}]
μ_0	Permeability of free space, $\mu_0 = 4\pi\ 10^{-7}$ H m^{-1}	[H m^{-1}]
μ_r	Relative permeability	
ν	Frequency deviation, $\nu = \omega/\omega_0 - \omega_0/\omega$	
ϱ	Radius in cylindrical coordinates (together with φ and z)	[m]
ϱ, $\underline{\varrho}$	Charge density and complex amplitude	[C m^{-3}]
ϱ^P, $\underline{\varrho}^\mathrm{P}$	Polarization charge density and complex amplitude	[C m^{-3}]
σ, $\underline{\sigma}$	Surface charge density and complex amplitude	[C m^{-2}]
Φ, $\underline{\Phi}$	Electric scalar potential and complex amplitude	[V]
$\underline{\boldsymbol{\varphi}}$	Complex potential vector	[V]
φ	Azimuth angle in cylindrical and spherical coordinates	[rad]
χ	Skin factor describing the increase of the element size	
Ψ	Magnetic flux	[Vs]
ω	Angular frequency, $\omega = 2\pi f$	[s^{-1}]
ω_0	Angular resonance frequency, $\omega_0 = 2\pi f_0$	[s^{-1}]
ω_SRF	Angular self-resonant frequency of a coil, $\omega_\mathrm{SRF} = 2\pi f_\mathrm{SRF}$	[s^{-1}]

Roman Letters

Notation	Description	Unit
$\mathbf{A}$	System matrix of the linear system $\mathbf{A}\mathbf{x} = \mathbf{b}$	
$\vec{A}, \underline{\vec{A}}$	Magnetic vector potential and complex amplitude	[Vs m^{-1}]
A	Area in $\mathbb{R}^3$	[m^2]
a	Radius of a circular conductor	[m]
$\mathbf{B}$	Sparse nodal connectivity matrix	
$\vec{B}, \underline{\vec{B}}$	Magnetic flux density and complex amplitude	[Vs m^{-2}]
$\mathbf{b}$	Excitation vector of the linear system $\mathbf{A}\mathbf{x} = \mathbf{b}$	
$\mathbf{C}_\mathrm{s}$	Dense short circuit capacitance matrix	[F]
C	Capacitance	[F]
c_0	Speed of light in free space, $c_0 = 299\,792\,458\,\mathrm{ms}^{-1}$	[ms^{-1}]
curl	Curl operator, $\operatorname{curl}\vec{F} = \nabla \times \vec{F}$	
$\mathbf{D}$	Sparse node reduction incidence matrix	
$\vec{D}, \underline{\vec{D}}$	Electric flux density and complex amplitude	[C m^{-2}]
$\mathrm{d}A, \mathrm{d}\vec{A}$	Infinitesimal area element, scalar and oriented	
$\mathrm{d}s, \mathrm{d}\vec{s}$	Infinitesimal path element, scalar and oriented	
$\mathrm{d}V$	Infinitesimal volume element	
d	Dissipation factor of a resonance circuit, reciprocal of Q_0	
div	Divergence operator, $\operatorname{div}\vec{F} = \nabla \cdot \vec{F}$	
$\vec{E}, \underline{\vec{E}}$	Electric field strength and complex amplitude	[V m^{-1}]
$\vec{e}_n$	Unit vector in $\mathbb{R}^3$ oriented in n-direction	
$\vec{F}$	Auxiliary vector field	
$\underline{f}$	Objective function	
f	Frequency	[Hz]
f_0	Resonance or working frequency	[Hz]
f_SRF	Self-resonant frequency of a coil	[Hz]
$G(\vec{r},\vec{r}\,')$	Green's function	[m^{-1}]
grad	Gradient operator, $\operatorname{grad}\Theta = \nabla\Theta$	
$\vec{H}, \underline{\vec{H}}$	Magnetic field strength and complex amplitude	[A m^{-1}]
h	Thickness of the substrate	[m]
$I, \underline{I}$	Current and complex amplitude	[A]
Im	Imaginary part of a complex number	
$\mathbf{i}$	Complex current vector	[A]
i	Index, $i \in \mathbb{N}$	
$\vec{J}, \underline{\vec{J}}$	Current density and complex amplitude	[A m^{-2}]
$\vec{J}^\mathrm{M}, \underline{\vec{J}}^\mathrm{M}$	Magnetization current density and complex amplitude	[A m^{-2}]
$\vec{J}^\mathrm{P}, \underline{\vec{J}}^\mathrm{P}$	Polarization current density and complex amplitude	[A m^{-2}]

Roman Letters

Notation	Description	Unit
j	Imaginary unit, $j = \sqrt{-1}$	
$\mathbf{K}$	Sparse terminal incidence matrix	
k	Wave number $k = \omega/c_0$	[m^{-1}]
k	Coupling factor between two conductors/coils	
$\mathbf{L}$	Dense matrix of partial inductances	[H]
L'	Per-unit-length inductance of a conductor $L' = L/l$	[H m^{-1}]
L	Inductance	[H]
L_{ext}	External inductance of a conductor/coil	[H]
L_{int}	Internal inductance of a conductor/coil	[H]
L_{mn}	Mutual inductance of conductors/coils m and n	[H]
l	Length of a conductor	[m]
l_{c}	Accumulated center length of a conductive trace	[m]
l_{i}	Accumulated inner length of a conductive trace	[m]
lim	Limit of a function or a sequence	
l_{o}	Accumulated outer length of a conductive trace	[m]
l_x	Outer length of rectangular spiral coil in x-direction	[m]
l_y	Outer length of rectangular spiral coil in y-direction	[m]
$\mathbf{M}$	Sparse mesh current incidence matrix	
$\vec{M}, \underline{\vec{M}}$	Magnetization and complex amplitude	[A m^{-1}]
M	Mutual inductance of two inductors	[H]
$\vec{m}_n$	n-th current basis function	[m^{-2}]
m	Index, $m \in \mathbb{N}$	
max	Maximum	
N	Number of elements	
$N_{\text{b}}, N_{\text{n}}, N_{\text{p}}$	Number of branches, nodes and panels in a PEEC system	
N_{c}	Number of corners of a spiral coil with $N_{\text{c}} = 4 N_{\text{turn}}$	
N_{d}	Number of design parameters	
N_t	Number of subdivisions of a rectangular conductor in t-direction	
N_{turn}	Number of turns of a spiral coil	
N_w	Number of subdivisions of a rectangular conductor in w-direction	
$\vec{n}$	Normal vector in $\mathbb{R}^3$	
n	Index, $n \in \mathbb{N}$	
O	Landau symbol representing the algorithmic complexity	
$\mathbf{P}$	Dense matrix of partial coefficients of potential	[F^{-1}]
$\vec{P}, \underline{\vec{P}}$	Electric polarization and complex amplitude	[C m^{-2}]
$\underline{P}$	Complex power	[W]
P	Coefficient of potential, indexed for multiconductor systems	[F^{-1}]
P_{l}	Ohmic losses in a resistive region due to a current flow	[W]
$\mathbf{p}$	Vector of N_{d} design parameters p_n	

Roman Letters

Notation	Description	Unit
p_n	n-th design parameter	
Q	Quality factor of a passive device	
Q, $\underline{Q}$	Charge and complex amplitude	[C]
Q_0	Quality factor of a resonance circuit	
Q_L	Intrinsic quality factor of a coil, $Q_L = X/R$	
$\underline{\mathbf{q}}$	Complex charge vector	[C]
q	Index, $q \in \mathbb{N}$	
$\mathbf{R}$	Diagonal matrix of partial resistances	[Ω]
R'	Per-unit-length resistance, $R' = R/l$	[$\Omega\,\text{m}^{-1}$]
R	Resistance	[Ω]
R_{DC}	DC Resistance	[Ω]
$\tilde{R}_{\text{DC}}$	Fractional DC resistance of a rectangular conductor bend	
Re	Real part of a complex number	
R_{S}	Radiation resistance	[Ω]
$\vec{r}$	Position vector in $\mathbb{R}^3$, observation point	[m]
$\vec{r}\,'$	Position vector in $\mathbb{R}^3$, source point	[m]
r	Radius in spherical coordinates (together with ϑ and φ)	[m]
r_0	Radius of a circular loop antenna	[m]
$\vec{S}$, $\underline{\vec{S}}$	Poynting vector and complex amplitude	[W m^{-2}]
S	Surface in $\mathbb{R}^3$	[m^2]
s	Spacing or pitch between two parallel conductors	[m]
t	Time	[s]
t	Thickness of a rectangular conductor	[m]
$\tan\delta$	Loss tangent of a dielectric material	
U	Voltage	[V]
$\underline{\mathbf{u}}$	Complex voltage vector	[V]
V	Volume in $\mathbb{R}^3$	[m^3]
v_q	q-th charge basis function	[m^{-3}]
W_{e}	Electric energy	[J]
W_{m}	Magnetic energy	[J]
w	Width of a rectangular conductor	[m]
w_{e}	Electric energy density	[J m^{-3}]
w_{m}	Magnetic energy density	[J m^{-3}]
X	Reactance	[Ω]
$\mathbf{x}$	Vector of state variables of the linear system $\mathbf{A}\mathbf{x} = \mathbf{b}$	
$\hat{\mathbf{x}}$	Vector of state variables of the adjoint linear system	
x	Cartesian coordinate	[m]
$\underline{Y}$	Admittance, $\underline{Y} = \underline{Z}^{-1}$	[S]
y	Cartesian coordinate	[m]

Roman Letters

Notation	Description	Unit
$\underline{Z}$	Impedance, $\underline{Z} = R + jX$	[Ω]
$\underline{Z}_R$	Reflected impedance of a remotely powered circuit	[Ω]
Z_0	Wave impedance of free space $Z_0 = \sqrt{\mu_0}/\sqrt{\varepsilon_0} \approx 377\,\Omega$	[Ω]
z	Cartesian coordinate	[m]

Bibliography

[1] N. Tesla, "Apparatus for transmitting electrical energy," U.S. Patent 1,119,732, Dec. 1, 1914.

[2] S. S. Mohammed, K. Ramasamy, and T. Shanmuganantham, "Wireless power transmission – a next generation power transmission system," *Int. Journal of Computer Applications*, vol. 1, no. 13, pp. 100–103, 2010. Online, [Jan. 2011]: http://www.ijcaonline.org/archives/volume1/number13/274-434

[3] A. Karalis, J. Joannopoulos, and M. Soljacic, "Efficient wireless non-radiative mid-range energy transfer," *Annals of Physics*, vol. 323, no. 1, pp. 34–48, 2008.

[4] G. E. Leyh and M. D. Kennan, "Efficient wireless transmission of power using resonators with coupled electric fields," in *40th North American Power Symp. (NAPS '08)*, Sep. 2008, pp. 1–4.

[5] A. Kurs, A. Karalis, R. Moffatt, J. D. Joannopoulos, P. Fisher, and M. Soljacic, "Wireless power transfer via strongly coupled magnetic resonances," *Sciencexpress*, Jun. 2007.

[6] M. J. Bueker, C. Reinhold, P. Scholz, U. Hilleringmann, T. Mager, and C. Hedayat, "Efficiency and field emission improvements of wireless energy transfer systems," in *Proc. 12th Sophia Antipolis MicroElectronics Conf. (SAME 2009)*, Sep. 2009. Online, [Jan. 2011]: http://same-conference.org/same_2009/documents/Demo_u_Booth/Demo_8.pdf

[7] M. Schwarz, R. Hauschild, B. Hosticka, J. Huppertz, T. Kneip, S. Kolnsberg, W. Mokwa, and H. Trieu, "Single chip CMOS image sensors for a retina implant system," in *Proc. IEEE Int. Symp. on Circuits and Systems (ISCAS'98)*, vol. 6, May 1998, pp. 645–648.

[8] H. M. Lu, C. Goldsmith, L. Cauller, and J.-B. Lee, "MEMS-based inductively coupled RFID transponder for implantable wireless sensor applications," *IEEE Trans. Magn.*, vol. 43, no. 6, pp. 2412–2414, Jun. 2007.

[9] K. Finkenzeller, *RFID Handbuch: Grundlagen und praktische Anwendungen induktiver Funkanlagen, Transponder und kontaktloser Chipkarten*, 4th ed. Hanser Fachbuchverlag, Aug. 2006. Online, [Jan. 2011]: http://www.rfid-handbook.com/

[10] W. John, P. Scholz, C. Reinhold, and G. Stönner, "Papierinfoträger haben bald ausgedient," *Funkschau*, no. 3, pp. 39–41, Mar. 2007. Online, [Jan. 2011]: http://www.funkschau.de/fileadmin/media/heftarchiv/articles/Jahrgang_2007/03_2007/fs_0703_s39-s41%20RFID.pdf

[11] J. C. Schuder, "Powering an artificial heart: Birth of the inductively coupled-radio frequency system in 1960," *Artificial Organs*, vol. 26, no. 11, pp. 909–915, 2002.

[12] E. S. Hochmair, "System optimization for improved accuracy in transcutaneous signal and power transmission," *IEEE Trans. Biomed. Eng.*, vol. BME-31, no. 2, pp. 177–186, Feb. 1984.

[13] S. Atluri and M. Ghovanloo, "A wideband power-efficient inductive wireless link for implantable microelectronic devices using multiple carriers," in *Proc. IEEE Int. Symp. on Circuits and Systems (ISCAS'06)*, May 2006, pp. 1131–1134.

[14] C. Fernandez, O. Garcia, R. Prieto, J. Cobos, S. Gabriels, and G. V. D. Borght, "Design issues of a core-less transformer for a contact-less application," in *Proc. IEEE Applied Power Electronics Conf. and Exposition (APEC'02)*, vol. 1, 2002, pp. 339–345.

[15] U.-M. Jow and M. Ghovanloo, "Design and optimization of printed spiral coils for efficient transcutaneous inductive power transmission," *IEEE Trans. Biomed. Circuits Syst.*, vol. 1, no. 3, pp. 193–202, Sep. 2007.

[16] S. Tang, S. Hui, and H. S.-H. Chung, "Coreless planar printed-circuit-coard (PCB) transformers–a fundamental concept for signal and energy transfer," *IEEE Trans. Power Electron.*, vol. 15, no. 5, pp. 931–941, Sep. 2000.

[17] M. Tiebout, *Low Power VCO Design in CMOS*, ser. Springer Series in Advanced Microelectronics. Springer Berlin Heidelberg, 2006, vol. 20.

[18] M. T. Thompson, "Inductance calculation technique – Part II: Approximations and handbook methods," *Power Control and Intelligent Motion*, pp. 1–11, Dec. 1999.

[19] W. G. Hurley and M. C. Duffy, "Calculation of self and mutual impedances in planar magnetic structures," *IEEE Trans. Magn.*, vol. 31, no. 4, pp. 2416–2422, Jul. 1995.

[20] H. A. Wheeler, "Inductance formulas for circular and square coils," in *Proc. IEEE*, vol. 70, no. 12, Dec. 1982, pp. 1449–1450.

[21] S. S. Mohan, M. del Mar Hershenson, S. P. Boyd, and T. H. Lee, "Simple accurate expressions for planar spiral inductances," *IEEE J. Solid-State Circuits*, vol. 34, no. 10, pp. 1419–1424, Oct. 1999.

[22] S. V. Kochetov, "Time- and frequency-domain modeling of passive interconnection structures in field and circuit analysis," Ph.D. dissertation, Otto-von-Guericke Universität Magdeburg, Sep. 2008. Online, [Jan. 2011]: http://diglib.uni-magdeburg.de/Dissertationen/2008/serkochetov.htm

[23] X. Hu, "Full-wave analysis of large conductor systems over substrate," Ph.D. dissertation, Massachusetts Institute of Technology, Jan. 2006. Online, [Jan. 2011]: http://hdl.handle.net/1721.1/35597

[24] R. Coccioli, T. Itoh, G. Pelosi, and P. P. Silvester, "Finite-element methods in microwaves: A selected bibliography," *IEEE Antennas Propag. Mag.*, vol. 38, no. 6, pp. 34–48, Dec. 1996.

[25] T. Weiland, "Eine Methode zur Lösung der Maxwellschen Gleichungen für sechskomponentige Felder auf diskreter Basis," *Electronics and Communication (AEÜ)*, vol. 31, no. 3, pp. 116–120, Mar. 1977.

[26] R. F. Harrington, *Field Computation by Moment Methods*. Robert E. Krieger Publishing Company Malaba, Florida, 1968.

[27] C. A. Brebbia and J. Dominguez, "Boundary element methods for potential problems," *Applied Mathematical Modelling*, vol. 1, no. 7, pp. 372–378, 1977.

[28] A. E. Ruehli, "Equivalent circuit models for three-dimensional multiconductor systems," *IEEE Trans. Microw. Theory Tech.*, vol. MTT-22, no. 3, pp. 216–221, Mar. 1974.

[29] CST - Computer Simulation Technology AG, Bad Nauheimer Str. 19, D-64289 Darmstadt, Germany. Online, [Jan. 2011]: http://www.cst.com

[30] M. Kamon, M. J. Tsuk, and J. K. White, "FastHenry: A multipole-accelerated 3-d inductance extraction program," *IEEE Trans. Microw. Theory Tech.*, vol. 42, no. 9, pp. 1750–1758, Sep. 1994.

[31] J. D. Jackson, *Classical Electrodynamics*, 3rd ed. John Wiley and Sons, 1998.

[32] G. Lehner, *Elektromagnetische Feldtheorie für Ingenieure und Physiker*, 5th ed. Springer, Berlin, 2005.

[33] H. A. Haus and J. R. Melcher, *Electromagnetic Fields and Energy*. Prentice Hall, Sep. 1989. Online, [Jan. 2011]: http://web.mit.edu/6.013_book/www/

Bibliography

[34] M. Wilke, "Zur numerischen Berechnung quasistationärer elektromagnetischer Felder im Zeitbereich," Ph.D. dissertation, Technische Universität Darmstadt, 2004.

[35] K. Meetz and W. L. Engl, *Elektromagnetische Felder: Mathematische und physikalische Grundlagen*. Springer, Berlin, 1980.

[36] H. K. Dirks, "Quasi-stationary fields for microelectronic applications," *Electrical Engineering*, vol. 79, no. 2, pp. 145–155, Apr. 1996.

[37] J. Larsson, "Electromagnetics from a quasistatic perspective," *Am. J. Phys.*, vol. 75, no. 3, pp. 230–239, Mar. 2007.

[38] K. Küpfmüller, W. Mathis, and A. Reibiger, *Theoretische Elektrotechnik Eine Einführung*, 18th ed. Springer Berlin Heidelberg, 2008.

[39] E. Baum and O. Erb, "Eddy currents and electrical surface charges," *Int. J. Numer. Modell. Electron. Networks Devices Fields*, vol. 16, pp. 199–218, May/Jun. 2003.

[40] C. G. Darwin, "The dynamical motions of charged particles," *Philosophical Magazine Series 6*, vol. 39, pp. 537–551, 1920.

[41] A. E. Ruehli and A. C. Cangellaris, "Progress in the methodologies for the electrical modeling of interconnects and electronic packages," *Proc. IEEE*, vol. 89, no. 5, pp. 740–771, May 2001.

[42] G. Antonini and A. E. Ruehli, "Fast multipole and multifunction PEEC methods," *IEEE Trans. Mobile Comput.*, vol. 2, no. 4, pp. 288–298, Oct.–Dec. 2003.

[43] M. Kamon, "Fast parasitic extraction qnd simulation of three-dimensional interconnect via quasistatic analysis," Ph.D. dissertation, Massachusetts Institute of Technology, Feb. 1998. Online, [Jan. 2011]: http://hdl.handle.net/1721.1/10048

[44] Z. Zhu, B. Song, and J. K. White, "Algorithms in fastimp: A fast and wideband impedance extraction program for complicated 3-d geometries," *IEEE Trans. Comput.-Aided Design Integr. Circuits Syst.*, vol. 24, no. 7, pp. 981–998, Jul. 2005.

[45] K. Nabors and J. White, "Multipole-accelerated capacitance extraction algorithms for 3-d structures with multiple dielectrics," *IEEE Trans. Circuits Syst. I*, vol. 39, no. 11, pp. 946–954, Nov. 1992.

[46] P. D. Patel, "Calculation of capacitance coefficients for a system of irregular finite conductors on a dielectric sheet," *IEEE Trans. Microw. Theory Tech.*, vol. 19, no. 11, pp. 862–869, Nov. 1971.

[47] V. Ardon, J. Aimé, O. Chadebec, E. Clavel, and E. Vialardi, "MoM and PEEC method to reach a complete equivalent circuit of a static converter," in *Proc. 20th Int. Zurich Symp. on Electromagnetic Compatibility (EMC)*, Jan. 2009, pp. 273–276.

[48] D. M. Pozar, *Microwave Engineering*, 3rd ed. John Wiley & Sons, 2005.

[49] D. H. Werner, "An exact integration procedure for vector potentials of thin circular loop antennas," *IEEE Trans. Antennas Propag.*, vol. 44, no. 2, pp. 157–165, Feb. 1996.

[50] K. Küpfmüller, *Einführung in die theoretische Elektrotechnik*, 13th ed. Springer-Verlag, Berlin, 1990.

[51] H. Haas, "Ein Beitrag zur Berechnung der Selbstinduktivität eines Torus," *Archiv für Elektrotechnik*, vol. 58, no. 6, pp. 305–308, Nov. 1976.

[52] A. M. Niknejad, "Analysis, simulation, and applications of passive devices on conductive substrates," Ph.D. dissertation, University of California at Berkeley, 2000. Online, [Jan. 2011]: http://www.eecs.berkeley.edu/~niknejad/pdf/NiknejadPhD.pdf

[53] H. M. Greenhouse, "Design of planar rectangular microelectronic inductors," *IEEE Trans. Parts, Hybrids, Packag.*, vol. PHP-10, no. 2, pp. 101–109, Jun. 1974.

[54] F. W. Grover, *Inductance Calculations*. D. Van Nostrand Co., New York, 1946; reprinted by Dover Publications, New York, 2004.

[55] P. Scholz, W. Ackermann, T. Weiland, and C. Reinhold, "Antenna modeling for inductive RFID applications using the partial element equivalent circuit method," *IEEE Trans. Magn.*, vol. 46, no. 8, pp. 2967–2970, Aug. 2010.

[56] S. Musunuri and P. L. Chapman, "Multi-layer spiral inductor design for monolithic DC-DC converters," in *Proc. Industry Applications Conf.*, vol. 2, Oct. 2003, pp. 1270–1275.

[57] J. Zheng, Y.-C. Hahm, V. K. Tripathi, and A. Weisshaar, "CAD-oriented equivalent-circuit modeling of on-chip interconnects on lossy silicon substrate," *IEEE Trans. Microw. Theory Tech.*, vol. 48, no. 9, pp. 1443–1451, Sep. 2000.

[58] A. C. Watson, D. Melendy, P. Francis, K. Hwang, and A. Weisshaar, "A comprehensive compact-modeling methodology for spiral inductors in silicon-based RFICs," *IEEE Trans. Microw. Theory Tech.*, vol. 52, no. 3, pp. 849–857, Mar. 2004.

[59] C. A. Chang, S.-P. Tseng, J. Y. Chuang, S.-S. Jiang, and J. A. Yeh, "Characterization of spiral inductors with patterned floating structures," *IEEE Trans. Microw. Theory Tech.*, vol. 52, no. 5, pp. 1375–1381, May 2004.

[60] Y. Cao, R. A. Groves, X. Huang, N. D. Zamdmer, J.-O. Plouchart, R. A. Wachnik, T.-J. King, and C. Hu, "Frequency-independent equivalent-circuit model for on-chip spiral inductors," *IEEE J. Solid-State Circuits*, vol. 38, no. 3, pp. 419–426, Mar. 2003.

[61] G. Antonini, D. Deschrijver, and T. Dhaene, "Broadband macromodels for retarded partial element equivalent circuit (rPEEC) method," *IEEE Trans. Electromagn. Compat.*, vol. 49, no. 1, pp. 35–48, Feb. 2007.

[62] A. Ahmad and P. Auriol, "Dielectric losses in power transformers under high frequency transients," in *Proc. IEEE Int. Symp. on Electrical Insulation*, Jun. 1992, pp. 460–463.

[63] H. A. Wheeler, "Formulas for the skin effect," in *Proc. IRE*, vol. 30, no. 9, Sep. 1942, pp. 412–424.

[64] C.-S. Yen, Z. Fazarinc, and R. L. Wheeler, "Time-domain skin-effect model for transient analysis of lossy transmission lines," in *Proc. IEEE*, vol. 70, no. 7, Jul. 1982, pp. 750–757.

[65] S. Kim and D. P. Neikirk, "Compact equivalent circuit model for the skin effect," in *Proc. IEEE MTT-S Int. Microwave Symp. Digest*, vol. 3, 1996, pp. 1815–1818.

[66] A. Görisch, "Netzwerkorientierte Modellierung und Simulation elektrischer Verbindungsstrukturen mit der Methode der partiellen Elemente," Ph.D. dissertation, Otto-von-Guericke-Universität Magdeburg, 2002.

[67] E. C. Levi, "Complex-curve fitting," *IRE Trans. Autom. Control*, vol. AC-4, pp. 37–44, 1959.

[68] G. Wunsch, *Theorie und Anwendung linearer Netzwerke, Teil 1, Analyse und Synthese.* Akademische Verlagsgesellschaft Geest & Portig K.-G., 1961.

[69] S. Kolnsberg, "Drahtlose Signal- und Energieübertragung mit Hilfe von Hochfrequenztechnik in CMOS-Sensorsystemen," Ph.D. dissertation, Gerhard-Mercator-Universität - Gesamthochschule Duisburg, Apr. 2001.

[70] C. Reinhold, P. Scholz, W. John, and U. Hilleringmann, "Efficient antenna design of inductive coupled RFID-systems with high power demand," *Journal of Communications*, vol. 2, no. 6, pp. 14–23, Nov. 2007. Online, [Jan. 2011]: http://www.academypublisher.com/jcm/vol02/no06/jcm02061423.pdf

[71] P. Scholz, C. Reinhold, W. John, and U. Hilleringmann, "Analysis of energy transmission for inductive coupled RFID tags," in *Proc. IEEE Int. Conf. on RFID*, Mar. 2007, pp. 183–190.

[72] A. E. Ruehli, "Inductance calculations in a complex integrated circuit environment," *IBM J. Res. Dev.*, vol. 16, no. 1, pp. 470–481, Sep. 1972.

[73] A. E. Ruehli and P. A. Brennan, "Efficient capacitance calculations for three-dimensional multiconductor systems," *IEEE Trans. Microw. Theory Tech.*, vol. 21, no. 2, pp. 76–82, Feb. 1973.

[74] W. Wessel, "Über den Einfluß des Verschiebungsstromes auf den Wechselstromwiderstand einfacher Schwingungskreise," *Annalen der Physik*, vol. 28, no. 5, pp. 59–70, 1937.

[75] H. Heeb and A. E. Ruehli, "Three-dimensional interconnect analysis using partial element equivalent circuits," *IEEE Trans. Circuits Syst. I*, vol. 39, no. 11, pp. 974–982, Nov. 1992.

[76] A. E. Ruehli and H. Heeb, "Circuit models for three-dimensional geometries including dielectrics," *IEEE Trans. Microw. Theory Tech.*, vol. 40, no. 7, pp. 1507–1516, Jul. 1992.

[77] G. Antonini, M. Sabatini, and G. Miscione, "PEEC modeling of linear magnetic materials," in *Proc. IEEE Int. Symp. on Electromagnetic Compatibility (EMC'06)*, vol. 1, 2006, pp. 93–98.

[78] M. E. Verbeek, "Partial element equivalent circuit (PEEC) models for on-chip passives and interconnects," *Int. J. Numer. Modell. Electron. Networks Devices Fields*, vol. 17, no. 1, pp. 61–84, Jan. 2004.

[79] A. Rong and A. C. Cangellaris, "Generalized PEEC models for three-dimensional interconnect structures and integrated passives of arbitrary shapes," in *Electrical Performance of Electronic Packaging*, Oct. 2001, pp. 225–228.

[80] V. Jandhyala, Y. Wang, D. Gope, and R. Shi, "Coupled electromagnetic-circuit simulation of arbitrarily-shaped conducting structures using triangular meshes," in *Int. Symp. on Quality Electronic Design*, 2002, pp. 38–42.

[81] A. E. Ruehli, G. Antonini, J. Esch, J. Ekman, A. Mayo, and A. Orlandi, "Nonorthogonal PEEC formulation for time- and frequency-domain EM and circuit modeling," *IEEE Trans. Electromagn. Compat.*, vol. 45, no. 2, pp. 167–176, May 2003.

[82] A. Müsing, J. Ekman, and J. W. Kolar, "Efficient calculation of non-orthogonal partial elements for the PEEC method," *IEEE Trans. Magn.*, vol. 45, no. 3, pp. 1140–1143, Mar. 2009.

[83] V. Vahrenholt, H.-D. Brüns, and H. Singer, "Razor blade functions in the PEEC method," in *Proc. IEEE Int. Symp. on Electromagnetic Compatibility (EMC'07)*, Jul. 9–13, 2007, pp. 1–6.

[84] L.-P. Schmidt, G. Schaller, and S. Martius, *Grundlagen der Elektrotechnik 3 - Netzwerke*, 1st ed.   Pearson Studium, 2006, vol. 3.

[85] M. Kamon, N. A. Marques, L. M. Silveira, and J. White, "Automatic generation of accurate circuit models of 3-d interconnect," *IEEE Trans. Compon., Packag., Manuf. Technol. B*, vol. 21, no. 3, pp. 225–240, Aug. 1998.

[86] F. Schmid, "Optimization of experimental computational electromagnetics code," Master's thesis, Lulea University of Technology, 2005.

[87] A. L. Zitzmann, "Fast and efficient methods for circuit-based automotive EMC simulation," Ph.D. dissertation, University of Erlangen-Nürnberg, 2007. Online, [Jan. 2011]: http://www.opus.ub.uni-erlangen.de/opus/volltexte/2007/544/

[88] C.-W. Ho, A. E. Ruehli, and P. A. Brennan, "The modified nodal approach to network analysis," *IEEE Trans. Circuits Syst.*, vol. 22, no. 6, pp. 504–509, Jun. 1975.

[89] J. Garrett, A. E. Ruehli, and C. Paul, "Recent improvements in PEEC modeling accuracy," in *Proc. IEEE Int. Symp. on Electromagnetic Compatibility*.   IEEE, Aug. 1997, pp. 347–352.

[90] J. E. Garrett, A. E. Ruehli, and C. R. Paul, "Accuracy and stability improvements of integral equation models using the partial element equivalent circuit (PEEC) approach," *IEEE Trans. Antennas Propag.*, vol. 46, no. 12, pp. 1824–1832, Dec. 1998.

[91] G. Antonini, A. Orlandi, and C. R. Paul, "Internal impedance of conductors of rectangular cross section," *IEEE Trans. Microw. Theory Tech.*, vol. 47, no. 7, pp. 979–985, Jul. 1999.

[92] W. T. Weeks, L. L. Wu, M. F. McAllister, and A. Singh, "Resistive and inductive skin effect an rectangular conductors," *IBM J. Res. Dev.*, vol. 23, no. 6, pp. 652–660, 1979.

[93] A. E. Ruehli, C. Paul, and J. Garrett, "Inductance calculations using partial inductances and macromodels," in *Proc. Int. Symp. on Electromagnetic Compatibility*, Aug. 1995, pp. 23–28.

[94] P. Silvester, "AC resistance and reactance of isolated rectangular conductors," *IEEE Trans. Power App. Syst.*, vol. PAS-86, no. 6, pp. 770–774, Jun. 1967.

[95] P. A. Brennan, N. Raver, and A. E. Ruehli, "Three-dimensional inductance computations with partial element equivalent circuits," *IBM J. Res. Dev.*, vol. 23, pp. 661–668, Nov. 1979.

[96] K. M. Coperich, A. E. Ruehli, and A. Cangellaris, "Enhanced skin effect for partial-element equivalent-circuit (PEEC) models," *IEEE Trans. Microw. Theory Tech.*, vol. 48, no. 9, pp. 1435–1442, Sep. 2000.

[97] M. Ouda and A. Sebak, "Efficient method for frequency dependent inductance and resistance calculations," in *Proc. IEEE WESCANEX Communications, Power and Computing Conf.*, vol. 2, May 1995, pp. 473–477.

[98] D. Gope, A. E. Ruehli, C. Yang, and V. Jandhyala, "(S)PEEC: time- and frequency-domain surface formulation for modeling conductors and dielectrics in combined circuit electromagnetic simulations," *IEEE Trans. Microw. Theory Tech.*, vol. 54, no. 6, pp. 2453–2464, Jun. 2006.

[99] K.-L. Krieger, "Integriertes Energie- und Datenübertragungssystem, Modelbildung und Systemorientierte Optimierung von Monolithisch Integrierten Mikrospulen," Ph.D. dissertation, Universität Bremen, 1999.

[100] G. Antonini, "PEEC modeling of magnetic materials and dispersive-lossy dielectrics," in *Workshop on PEEC Modeling*. Lulea University of Technology, Nov. 2007, pp. 1–59. Online, [Jan. 2011]: http://www.ltu.se/csee/research/eislab/news/1.32682

[101] F. Ling, J. Liu, and J.-M. Jin, "Efficient electromagnetic modeling of three-dimensional multilayer microstrip antennas and circuits," *IEEE Trans. Microw. Theory Tech.*, vol. 50, no. 6, pp. 1628–1635, Jun. 2002.

[102] S. M. Rao, T. K. Sarkar, and R. F. Harrington, "The electrostatic field of conducting bodies in multiple dielectric media," *IEEE Trans. Microw. Theory Tech.*, vol. 32, no. 11, pp. 1441–1448, Nov. 1984.

Bibliography

[103] J. R. Phillips, "Rapid solution of potential integral equations in complicated 3-dimensional geometries," Ph.D. dissertation, Massachusetts Institute of Technology, Jun. 1997. Online, [Jan. 2011]: http://hdl.handle.net/1721.1/43400

[104] M. Gimignani, A. Musolino, and M. Raugi, "Integral formulation for nonlinear magnetostatic and eddy currents analysis," *IEEE Trans. Magn.*, vol. 30, no. 5, pp. 3024–3027, Sep. 1994.

[105] J.-P. Keradec, E. Clavel, J.-P. Gonnet, and V. Mazauric, "Introducing linear magnetic materials in PEEC simulations. Principles, academic and industrial applications," in *Proc. Industry Applications Conf.*, vol. 3, Oct. 2005, pp. 2236–2240.

[106] X. Zhang, W. H. Chen, and Z. Feng, "Novel SPICE compatible partial-element equivalent-circuit model for 3-d structures," *IEEE Trans. Microw. Theory Tech.*, vol. 57, no. 11, pp. 2808–2815, Nov. 2009.

[107] A. Devgan, H. Ji, and W. Dai, "How to efficiently capture on-chip inductance effects: Introducing a new circuit element K," in *Proc. IEEE/ACM Int. Conf. on Computer Aided Design (ICCAD'00)*, 2000, pp. 150–155.

[108] H. Ji, A. Devgan, and W. Dai, "KSim: A stable and efficient RKC simulator for capturing on-chip inductance effect," in *Proc. Asia and South Pacific Design Automation Conf. (ASPDAC'01)*, Jun. 2001, pp. 379–384.

[109] G. Antonini, "Fast multipole method for time domain PEEC analysis," *IEEE Trans. Mobile Comput.*, vol. 2, no. 4, pp. 275–287, Oct. – Dec. 2003.

[110] N. K. Nikolova, J. W. Bandler, and M. H. Bakr, "Adjoint techniques for sensitivity analysis in high-frequency structure CAD," *IEEE Trans. Microw. Theory Tech.*, vol. 52, no. 1, pp. 403–419, Jan. 2004.

[111] B. D. H. Tellegen, "A general network theorem, with applications," *Philips Research Reports*, vol. 7, pp. 259–269, Aug. 1952.

[112] V. A. Monaco and P. Tiberio, "Computer-aided analysis of microwave circuits," *IEEE Trans. Microw. Theory Tech.*, vol. 22, no. 3, pp. 249–263, Mar. 1974.

[113] J. I. Toivanen, R. A. E. Mäkinen, S. Järvenpää, P. Ylä-Oijala, and J. Rahola, "Electromagnetic sensitivity analysis and shape optimization using method of moments and automatic differentiation," *IEEE Trans. Antennas Propag.*, vol. 57, no. 1, pp. 168–175, Jan. 2009.

[114] N. K. Nikolova, J. Zhu, D. Li, M. H. Bakr, and J. W. Bandler, "Sensitivity analysis of network parameters with electromagnetic frequency-domain simulators," *IEEE Trans. Microw. Theory Tech.*, vol. 54, no. 2, pp. 670–681, Feb. 2006.

[115] P. Scholz, W. Ackermann, and T. Weiland, "Derivatives of partial inductances for the sensitivity analysis in PEEC systems," in *Proc. 2010 URSI Int. Symp. on Electromagnetic Theory (EMTS)*, Aug. 2010, pp. 48–51.

[116] S. Amari, "Numerical cost of gradient computation within the method of moments and its reduction by means of a novel boundary-layer concept," in *Proc. IEEE MTT-S Int. Microwave Symp. Digest*, vol. 3, 2001, pp. 1945–1948.

[117] D. Meeker, "FEMM – Finite Element Method Magnetics." Online, [Jan. 2011]: http://www.femm.info

[118] J. D. Cockcroft, "Skin effect in rectangular conductors at high frequencies," *Proceedings of the Royal Society of London. Series A, Containing Papers of a Mathematical and Physical Character*, vol. 122, no. 790, pp. 533–542, 1929.

[119] L. J. Giacoletto, "Frequency- and time-domain analysis of skin effects," *IEEE Trans. Magn.*, vol. 32, no. 1, pp. 220–229, Jan. 1996.

[120] K. Jakubiuk and P. Zimny, "Skin effect in rectangular conductors," *Journal of Physics A: Mathematical and General*, vol. 9, no. 4, pp. 669–676, 1976.

[121] H.-G. Groß, "Die Berechnung der Stromverteilung in zylindrischen Leitern mit rechteckigem und elliptischem Querschnitt," *Archiv für Elektrotechnik*, vol. 34, no. 5, pp. 241–268, May 1940.

[122] L. N. Trefethen, "Analysis and design of polygonal resistors by conformal mapping," *Journal of Applied Mathematics and Physics (ZAMP)*, vol. 35, pp. 692–704, 1984.

[123] P. Scholz, W. Ackermann, and T. Weiland, "PEEC antenna modeling of rectangular spiral inductors for RFID systems," *COMPEL*, vol. 29, no. 6, pp. 1453–1463, Dec. 2010.

[124] J. M. López-Villegas, J. Samitier, C. Cané, P. Losantos, and J. Bausells, "Improvement of the quality factor of RF integrated inductors by layout optimization," *IEEE Trans. Microw. Theory Tech.*, vol. 48, no. 1, pp. 76–83, Jan. 2000.

[125] J. Sieiro, J. M. López-Villegas, J. Cabanillas, J. A. Osorio, and J. Samitier, "A physical frequency-dependent compact model for rf integrated inductors," *IEEE Trans. Microw. Theory Tech.*, vol. 50, no. 1, pp. 384–392, Jan. 2002.

[126] J. C. Lagarias, J. A. Reeds, M. H. Wright, and P. E. Wright, "Convergence properties of the nelder–mead simplex method in low dimensions," *SIAM Journal of Optimization*, vol. 9, no. 1, pp. 112–147, 1998.

Bibliography

[127] Agilent Technologies, *4294A Precision Impedance Analyzer, 40 Hz to 110 MHz*. Online, [Jan. 2011]: http://cp.literature.agilent.com/litweb/pdf/5968-3809E.pdf

[128] A. M. Niknejad, "Analysis, design, and optimization of spiral inductors and transformers for Si RF ICs," Master's thesis, University of California at Berkeley, 1997. Online, [Jan. 2011]: http://www.eecs.berkeley.edu/~niknejad/pdf/NiknejadMasters.pdf

[129] M. Enohnyaket, "PEEC modeling and verification for broadband analysis of air-core reactors," Master's thesis, Lulea University of Technology, 2007. Online, [Jan. 2011]: http://epubl.luth.se/1402-1757/2007/51/

[130] C. Hoer and C. Love, "Exact inductance equations for rectangular conductors with applications to more complicated geometries," *J. Res. Nat. Bur. Stand.*, vol. 69C, no. 2, pp. 127–137, Jan. 1965.

[131] G. Zhong and C.-K. Koh, "Exact closed-form formula for partial mutual inductances of rectangular conductors," *IEEE Trans. Circuits Syst. I*, vol. 50, no. 10, pp. 1349–1352, Oct. 2003.

[132] K.-Y. Su and J.-T. Kuo, "Analytical evaluation of inductance of spiral inductors using partial element equivalent circuit (PEEC) technique," in *Proc. IEEE Antennas and Propagation Society Int. Symp.*, vol. 1, Jun. 2002, pp. 364–367.

[133] F. F. Martens, "Über die gegenseitige Induktion und ponderomotorische Kraft zwischen zwei stromdurchflossenen Rechtecken," *Annalen der Physik*, vol. 334, no. 10, pp. 959–970, 1909.

[134] J.-T. Kuo, K.-Y. Su, T.-Y. Liu, H.-H. Chen, and S.-J. Chung, "Analytical calculation for DC inductances of rectangular spiral inductors with finite metal thickness in the PEEC formulation," *IEEE Microw. Wireless Compon. Lett.*, vol. 16, no. 2, pp. 69–71, Feb. 2006.

[135] A. B. Birtles and B. J. Mayo, "Comments on "Calculation of capacitance coefficients for a system of irregular finite conductors on a dielectric sheet"," *IEEE Trans. Microw. Theory Tech.*, vol. 21, no. 8, pp. 568–569, Aug. 1973.

[136] T. F. Eibert and V. Hansen, "On the calculation of potential integrals for linear source distributions on triangular domains," *IEEE Trans. Antennas Propag.*, vol. 43, no. 12, pp. 1499–1502, Dec. 1995.

[137] P. Balaban, "Calculation of the capacitance coefficients of planar conductors on a dielectric surface," *IEEE Trans. Circuit Theory*, vol. 20, no. 6, pp. 725–731, Nov. 1973.

[138] R. Bancroft, "A note on the moment method solution for the capacitance of a conducting flat plate," *IEEE Trans. Antennas Propag.*, vol. 45, no. 11, p. 1704, Nov. 1997.

[139] J.-T. Kuo and K.-Y. Su, "Analytical evaluation of the MoM matrix elements for the capacitance of a charged plate," *IEEE Trans. Microw. Theory Tech.*, vol. 50, no. 5, pp. 1435–1436, May 2002.

I want morebooks!

Buy your books fast and straightforward online - at one of world's fastest growing online book stores! Environmentally sound due to Print-on-Demand technologies.

Buy your books online at
www.morebooks.shop

Kaufen Sie Ihre Bücher schnell und unkompliziert online – auf einer der am schnellsten wachsenden Buchhandelsplattformen weltweit! Dank Print-On-Demand umwelt- und ressourcenschonend produziert.

Bücher schneller online kaufen
www.morebooks.shop

KS OmniScriptum Publishing
Brivibas gatve 197
LV-1039 Riga, Latvia
Telefax:+371 686 204 55

info@omniscriptum.com
www.omniscriptum.com

MIX
Papier aus verantwortungsvollen Quellen
Paper from responsible sources
FSC® C105338

Printed by Books on Demand GmbH, Norderstedt / Germany